Sharing the Harvest

The USDA Sustainable Agriculture Research and Education (SARE) Program works to increase knowledge about—and help farmers adopt—practices that are economically viable, environmentally sound and that enhance quality of life for farmers, rural communities, and society as a whole. Authorized in the 1985 Farm Bill, SARE began funding competitive grants in 1988. Since then, it has funded more than twelve hundred projects aimed at improving the sustainability of farming. Through newsletters, books, Web sites and other resources, SARE helps to spread information about innovative, practical, profitable and ecological approaches to agriculture.

This book—the result of partnerships with authors Elizabeth Henderson and Robyn Van En, and publisher Chelsea Green—reflects Northeast SARE's commitment to supporting food production, processing, and marketing systems that sustain rural communities and support agricultural activities in the region's rural, urban, and suburban areas.

Farmers throughout the country are under tremendous pressures—from suburban development driving up the costs of farmland to high prices for many resources they need to low prices for their products. In addition, the food we eat is often grown far away from the table—the typical produce is said to have traveled well over one thousand miles—and less fresh and lower-quality fruits and vegetables are often the result. The tremendous interest in Community Supported Agriculture (CSA) farms among farmers as well as the public is an indication that this approach has struck a chord with people who want high-quality fresh produce and also want to have a more direct connection with that most basic of human needs—food.

For more information, please contact SARE at Hills Building, University of Vermont, Burlington, VT 05405-0082. Phone: 802-656-0471; Fax: 802-656-4656; Email: nesare@zoo.uvm.edu. Web site: http://www.uvm.edu/~nesare/.

SHARING THE HARVEST

A Guide to Community Supported Agriculture

Elizabeth Henderson with Robyn Van En

Foreword by Joan Dye Gussow

CHELSEA GREEN PUBLISHING COMPANY
White River Junction, Vermont
Totnes, England

To Robyn, and to our sons, David and Andy, the next generation, who make it worthwhile to struggle for a peaceful and sustainable world.

Cover design by Ann Aspell.
Designed by Andrea Gray.

Printed in the United States
First printing, April 1999

02 01 00 99 1 2 3 4 5

Printed on acid-free, recycled paper.

Library of Congress Cataloging-in-Publication Data

Henderson, Elizabeth, 1943–
Sharing the harvest: a guide to community supported agriculture/
Elizabeth Henderson; with introduction by Robyn Van En; forward by Joan Dye Gussow.
p. cm.
Includes bibliographical references and index.
ISBN 1-890132-23-2 (alk. paper)
1. Collective farms—United States. 2. Community gardens—United States.
3. Agriculture, Cooperative—United States. 4. Farm produce—United States—Marketing.
5. Agriculture—Economic aspects—United States. I. Title.
HD1492.U6H46 1999
334' .683'0973—dc21 98-53019

Chelsea Green Publishing Company
Post Office Box 428
White River Junction, VT 05001
(800) 639-4099
http://www.chelseagreen.com

CONTENTS

FOREWORD

by Joan Dye Gussow

ACROSS THIS COUNTRY, A MOVEMENT is spreading that acknowledges a long-ignored reality: most of what we pay for our food goes to companies who transport, process, and market what comes off the farm, not to farmers themselves. The people who actually grow food don't get paid enough to keep on doing it. And so, from Maine to California, some farmers are being supported by voluntary communities of eaters organized to pay growers directly for what they produce. Bypassing the supermarket, the middlemen, and the 747s that fly produce around the world, these folks are getting fresh local produce in season, at reasonable prices. This is a book about eater communities who are buying what their local farmers grow, and this system is called—appropriately—Community Supported Agriculture.

If this is the first time you've heard of Community Supported Agriculture, you should turn immediately to the first chapter. There you can discover why various farmers and eaters are joining in these groups, and learn what some of them experience as they exchange wholesome food and financial security. If you have heard of CSAs but know little about them, then you need to study this recounting of the many different ways in which producers and consumers have come together to find mutually satisfactory solutions to their financial, social, and culinary needs. And finally, if you are an active member of CSA, you surely need to sit down with Liz Henderson, farmer extraordinary—and Robyn Van En, late great lady of the CSA movement—to learn how to solve the problems that will crop up (if they have not already done so) as you and your co-shares try to find ways of making good food available outside the dominant marketplace, for the long-term benefit of everyone.

I was two-thirds of the way through the manuscript of this book when I went to a meeting of core group members of several New York City CSAs. (Yes, there are people in the heart of the metropolis getting food directly from farmers.) As I listened to these pioneers seeking help from each other in resolving emerging dilemmas, I kept wishing they had this book. Under the major headings "Getting Started," "Getting Organized," "The Food," and "Many Models," they would have found ideas on everything from "How to Choose a Farmer" and "Acquiring Land," through "Farmer Earnings" and "Startup Expenses," to "Distributing the Harvest," "Regional Networking," and "Including Low-Income Members."

In listing these topics, I realize that I risk making this book sound like a dull instruction manual or a guidebook to CSAs. It is a guide, and it is instructive, but it is far more than either of those, and it is surely not dull. It is, in fact, a delight to read, since the woman who finally brought it together, after the original author succumbed to an unacceptably early death, is a strong and gifted writer. She recognizes other good writers when she reads them and quotes many of them when it is appropriate—including many of the participants in these enterprises.

You will learn a lot from this book. I know something about most of the subjects covered, but found myself highlighting the manuscript as I went, marking important facts or wonderful quotes. You will find hundreds of ideas here for dealing with issues that arise at each step along the way, but you will not find a single set of instructions for organizing or maintaining a CSA, since uniformity is not the movement's (or the authors') goal.

But if there is no single approach laid out in this book, there is, perhaps, a single goal. Namely, to encourage more people to join in the struggle to save farming. The authors are appropriately desperate to save farmers and farmland, and want more people to understand the importance of their passion. "I have searched particularly," Liz Henderson writes about her research for this unique compilation of experiences, "for ideas on how to entice, entrap, entangle, or engage as many people as possible in community support for local farming

Shortly after I had visited the early CSA begun by Robyn Van En, I met another person active in the movement who made a comment I have never forgotten. "Community Supported Agriculture," he asserted, "is a kind of barter relationship. You pay this money up front in the fall to help support a local farmer, and next summer you get all this free produce." As things have turned out, not all CSAs can or want to collect money up front, and not all sharers experience the produce as "free." While generosity and community are goals, this fascinating book makes clear that it's seldom simple to achieve them. In a society where free choice is a mantra and everything has been assigned a price, self-imposed limitations do not come easily and community doesn't build spontaneously.

If we hope to keep eating, however, we need to keep farmers in business; and if we want to keep farmers in business, it's time for all of us, ordinary citizens and policymakers alike, to begin learning how that might be done. *Sharing the Harvest* is a great place to start.

ACKNOWLEDGMENTS

ROBYN VAN EN SHOULD HAVE BEEN the author of this book. Her untimely death from asthma took her from the work she loved on behalf of Community Supported Agriculture, and from the completion of this text. I had volunteered to help out with a few chapters, and to give moral and technical support. When Robyn passed, I took on what I thought would be the job of polishing her first draft for publication. Unfortunately, I discovered that there was not enough to polish, so I have rewritten the book substantially, trying to keep all of Robyn's concepts, and as much of her prose as possible.

First of all, I want to thank the hundreds of CSA farmers and sharers all over the country who have contributed materials, ideas, suggestions, and enthusiasm for this book, and allowed me to tell their stories. It is truly an honor to be associated with this creative, hopeful movement.

I wish to express my appreciation to C. R. Lawn for reading and rereading the drafts, and doing everything in his power to make it at least as lively and concrete as a Fedco Seed Catalogue. His colleague at Fedco, Robin Sherman, suggested *Sharing the Harvest* as a title for the book. Thanks as well to Marianne Simmons, Portia Weiskel, Jennifer Bokaer-Smith, and Patricia Mannix for reading parts of the manuscript and giving helpful suggestions and encouragement. And special thanks to Beth Holtzman, NE SARE Communications Specialist, for guidance, for the chapter on the Intervale, and for shepherding this project through the SARE process. I am also grateful to Ben Watson for his knowledgeable and tactful editing on behalf of Chelsea Green.

This book would not have made it into print without the generous financial and moral support of Fred Magdoff and the Northeast Sustainable Agriculture Research and Education Program (NE SARE). The NE SARE Administrative Council accepted my proposal in 1995 to provide financial support to Robyn so that she could quit her two part-time jobs and concentrate on this book, then waited patiently as the writing got underway, and agreed to transfer the support to me, as Robyn's substitute. NE SARE has also invested jointly with Chelsea Green in the publication of the book.

I also want to give thanks to the many members

of the Genesee Valley Organic Community Supported Agriculture Project (GVOCSA), and especially to Alison Clarke, our fearless organizer, for joining in this great experiment in local food production with me; to Greg Palmer, for years of listening while hoeing, picking, and packing at my side; and to my former partner David Stern, from whom I learned a great deal about organic farming, including how risky it can be. My gratitude goes as well to Nancy Kasper for an artistic refuge in which to finish writing this book.

And finally, I would like to acknowledge the steadying presence of Harry Henderson, my father-in-law, close friend, and mentor in this writing business.

Any inaccuracies or mistakes are purely my own.

INTRODUCTION

by Robyn Van En (December 1996)

Whatever you can do or dream you can do—begin it. Boldness has genius, power, and magic in it. Begin it now.

—GOETHE

IN 1983, MY YOUNG SON DAVID AND I moved to Massachusetts from the Northern California redwood forests to experience New England's colored leaves of autumn and snowy winters, while I finished my training as a Waldorf kindergarten teacher. We planned to be in the area for a year, to get our bearings, and then find property to buy. I would look for a place with a bit of land for me to resume growing bouquet flowers and perennial stock for landscapers, as I had done in California while I went to school. Within a month after our arrival, circumstances motivated me to find the "place," and somewhat by accident, maybe more by providence, we landed at Indian Line Farm, in the village of South Egremont.

Indian Line Farm was so named almost two hundred years ago as one of the early Dutch settlements built at an imaginary line beyond which was Indian country. Supposedly, Johnny Appleseed came along Jugend Road, planting his apple trees. Shay's Rebellion, largely a farmers' uprising against unfair taxation, which influenced the writing of the Constitution, took place right over the hill.

Indian Line is a beautiful farm with a big, old house and a huge dairy barn resting upon table flat, fertile bottomland. Beneath the topsoil is limestone ledge, giving the soil a mellow pH. Nearby is an alkaline fen (wetland), a rare habitat, protected and monitored by the Nature Conservancy. The middle of the field looks out on Jugend Mountain, formerly an Indian observation point, with the remains of their long houses and another old farmstead in the state park at the foot. Over the top of the mountain runs a portion of the Appalachian Trail, leading into Connecticut, just behind the first Community Land Trust and the E. F. Schumacher Library.

I was raised in California and for years had cultivated a vegetable plot, besides the bouquet flowers and perennials, but to start managing a 60-acre, retired dairy farm in the Berkshire Hills was a huge departure for me.

Soon after I arrived in the neighborhood and joined the staple foods buying club, I had a conversation with Susan Witt from the E. F. Schumacher Society. We discussed what I would be growing at the farm the forthcoming spring, besides flowers, as I had a whole lot of available ground.

I found out that most of the buying club members had their own summer gardens, but went to distant farms or the supermarket for their winter vegetables. Why shouldn't I grow those storage crops? At a coop meeting, people said they would buy anything I grew, so I planted accordingly. By planting primarily potatoes, carrots, onions, garlic, and winter squash, I had fairly good returns, with my market ready and waiting, but I still carried all of the capitalization expenses, all of the work, and all of the risk.

I spent long periods, generally while hoeing, trying to formulate a better way to oblige both the grower and the eaters. The better way would be something cooperative; an arrangement that would allow people to draw upon their combined abilities, expertise, and resources for the mutual benefit of all concerned. It would also bring the people producing the food closer to the people who were eating the food, and the eaters closer to the land.

In the middle of my second growing season, as I pondered this agricultural conundrum, Jan Vandertuin visited the farm. He had recently returned to New England from Switzerland and was anxious to share the experience he'd had working with a couple of farmers there. These farmers had asked their regular customers to pay a share of the farm's annual production expenses in exchange for a weekly share of the produce. Shares of the vegetables, meat, and dairy products were made available to them. After talking only a few minutes, Jan and I knew that we should do the same at Indian Line Farm.

Jan Vandertuin, John Root, Jr. (co-director of Berkshire Village, a group home for handicapped adults, a stone's throw from the Schumacher Society and a mile from Indian Line), and I got busy organizing. We introduced the "share the costs to share the harvest" concept to the surrounding community by way of the Apple Project in the autumn of 1985. People paid in advance for family-sized shares of the apple harvest. Most of those folks signed up for the vegetable shares that we offered for the following spring. We were determined to make this happen, so we were very busy educating community members about the vegetable shares at the same time we were interviewing potential gardeners and farmers. No one had ever heard of being paid for vegetables in advance, before the first seed was planted, but we were finally approached by Hugh Ratcliffe, who started breaking ground that fall for the eventual spring planting. The rest of us carried on with the Apple Project and community education.

During the winter of 1985–86, we met each week to discuss and develop the logistics and procedures necessary to accomplish our goals: local food for local people at a fair price to them and a fair wage to the growers. The members' annual commitment to pay their share of the production costs and to share the risk as well as the bounty set this apart from any other agricultural initiative.

We didn't take any step of this process lightly. We discussed and debated long into the nights the necessary policies and procedures, besides the possible names for the project that would convey its full intent and purpose. We finally decided on Community Supported Agriculture, which could be transposed to Agriculture Supported Communities and say what we needed in the fewest words. CSA to ASC was the whole message. We knew it was a mouthful and doesn't fit easily into conversation or text, but to this day I can't think of a better way to name what it's all about. People have tried substituting other words ("consumer shared agriculture" is what some people call it in Canada), and I found obvious discomfort with the word "community" when I tried to explain the concept in the former Soviet Union. People have problems with "supported," too. Please know that each word was chosen after lengthy consideration. I personally was ada-

CLEMENS KALISCHER

Robyn Van En

a bag of vegetables twice a week throughout the growing season, and twice a month from the root cellar during the winter months. This proved to be too many vegetables for most households. The next year, many found a friend or neighbor to take the second bag, and by the third year, we pared it down to one bag a week. Larger households, restaurants, or markets bought multiple shares.

We have come a long way since that first season. As the founding core group, we learned a lot, and realized that there was a lot more to learn. Working with a group of people in a manner that honors and makes the best use of the collective expertise and resources, at the same time becoming familiar with and adjusting to different personalities and agendas, is not easy. The logic, simplicity, and earthly need for the concept carried us through. Despite our many differences, we created a working prototype and replicable model of CSA. After four years, we separated—a rather grizzly process—but, looking back, I can honestly say I wouldn't have learned nearly so much about myself, about group dynamics, about CSA's pitfalls and potential, and about the larger community, both locally and regionally, if the split had been easier.

mant about using the word "agriculture" rather than calling the project CS Farms, because I didn't want to exclude similar initiatives from taking place on a corner lot in downtown Boston. We had to call it something fast because the project was ready to go.

To secure the land for existing and potential members, I leased my garden site (approximately five acres) to the CSA, an unincorporated association, for three years with an option to buy the land into an agricultural trust after the third year.

We offered our first shares of the vegetable harvest in the spring of 1986. Early members received

I wrote the first edition of the CSA startup manual, the *Basic Formula to Create Community Supported Agriculture*, largely in self-defense, as I was spending hours in the field, on the phone, or in conference halls talking about our experiment at Indian Line Farm. I designed the manual to help readers answer many of their questions about CSA and to formulate questions specific to their own situations. Translating CSA to the North American landscape and mentality, which are vastly different in scale, available resources, and culture than Japan and Western Europe, was a challenge. CSA has certain fundamental logistical points that are similar no matter where or how it is practiced, but, at the same time, it is largely an evolving and highly adaptive process, as I tried to share and convey through the *Basic Formula* manual.

By spring of 1996, close to six hundred CSA projects engaging at least one hundred thousand people throughout the United States and Canada were putting seeds into the Earth. An armchair social study conducted by Jean-Pierre Schwartz—a founding member of the first Washington, D.C., area CSA, and a founding board member of CSA of North America (CSANA)—compared the current number of CSAs, their rate of increase to date, and their potential increase, to the boom of bed-and-breakfast establishments springing up across the country. From this, Schwartz concluded that we may see ten thousand CSA farms and gardens by the year 2000. I certainly hope so. That could mean an average of two hundred projects in each state and Canada with nearly two million people involved, signifying an important shift in mainstream consciousness. We would be well on our way toward a "trend," with 2 percent of the overall population becoming aware of this new alternative for purchasing fresh food. People don't even need to be actual CSA members, just marginally aware. If we can achieve that much consciousness in the U.S. and Canada, CSA's presence certainly will have increased around the world too.

While the Indian Line Farm project was modeled after Jan's experience in Switzerland, I have since learned that the CSA equivalent was developed first in Japan in 1965, initiated by a group of women concerned with the use of pesticides, the increase in processed and imported foods, and the corresponding decrease in the local farm population.

The group approached a local farmer and worked out the terms of their cooperative agreement, and the teikei movement was born. Literally translated, *teikei* means "partnership" or "cooperation," but according to teikei members in Japan, the more philosophical translation is "food with the farmer's face on it."

The philosophical aspects of teikei are also represented in the names of their groups, such as the Society for Reflecting the Throwaway Age, the Young Leaf Society, and the Society Protecting the Earth. This aspect is fairly common in North American CSA as well, with names like Walk Softly CSA, Twin Creek Shared Farm, Deliberate Living CSA, Gathering Together Farm, Caretaker Community Farm, and Heartbeet.

The teikei farms or garden sites in Japan tend to be quite small and intensively farmed. It is common for a group of farmers, dispersed throughout the countryside, to supply a large group of urban members by delivering their produce to one of the express train depots for transport to a central pickup point in the city. It is also common to have groups of fifteen hundred households networking with a group of fifteen farmers to get a consistent and diverse selection of products. Typically, members are supplied with their weekly vegetables, herbs, fruit, fish, poultry, eggs, grain, and soy product needs, along with soap and candles from a community-supported cottage industry. (For a more in-depth examination of teikei, see "Returning Relationships to Food: The Teikei Movement in Japan" by Annie Main and Jered Lawson, p. 214.)

In 1992, I had the great pleasure of cohosting members from one of the teikei groups that was travelling in the U.S. and Canada. They had come to see organic farms in general and some CSA projects specifically. I was amazed how similar each of our concerns and visions were for teikei and CSA and for the future of agriculture as the basis of all culture. It was truly a full circle experience for me.

Equally empowering to both the community and the farmers, CSA offers solutions to common problems facing farmers and communities worldwide. Ultimately, the concept is capable of engaging and empowering people to a capacity that has been all but lost in this "modern" world.

— PART I —

CSA in Context

WHAT IS COMMUNITY SUPPORTED AGRICULTURE? 1

You must be *the change you wish to see in the world.*

—Mahatma Gandhi

COMMUNITY SUPPORTED AGRICULTURE is a connection between a nearby farmer and the people who eat the food that the farmer produces. Robyn Van En summed it up as "food producers + food consumers + annual commitment to one another = CSA and untold possibilities." The essence of the relationship is the mutual commitment: the farm feeds the people, the people support the farm and share the inherent risks and potential bounty. Doesn't sound like anything very new—for most of human history, people have been connected with the land that fed them. Growing (or hunting and gathering) food somewhere nearby is basic to human existence, as basic as breathing, drinking, and sexual reproduction. If this basic connection breaks down, there is sure to be trouble.

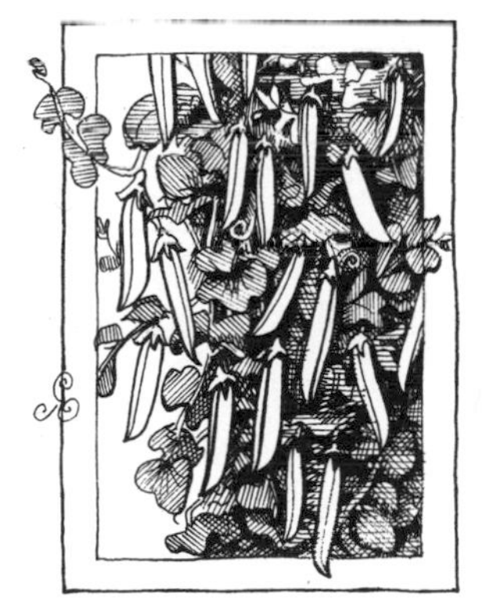

For the masses of people in the United States today, this connection *has* been broken. Most people do not know where or how their food is grown. They cannot touch the soil or talk to the farmer who tends it. Food comes from stores and restaurants and vending machines. It has been washed, processed, packaged, maybe even irradiated, and transported long distances. As trade becomes "free," the food travels even longer distances. Stores in the northeast used to carry Florida tomatoes in the winter. Today, with the North American Free Trade Agreement (NAFTA), the tomatoes come all the way from Mexico. After a few more years under the General Agreement on Tariffs and Trade (GATT), apples from China may crowd out the Washington State apples, which for the past decade have crowded the New York State apples off the supermarket bins in New York City. Farmers alone have been shouldering the risks of this increasingly ruthless global market, which has forced millions of them from the land. CSA offers one of the most hopeful alternatives.

Robyn Van En dedicated the last ten years of her much-too-short life to spreading the word about CSA, speaking at conferences and workshops, and giving advice and trouble-shooting in person or over the phone. She was the founder and sole staff person for CSA of North America (CSANA), which distributed the handbook she wrote in 1988, *Basic Formula to Create Community Supported Agriculture*. A common cliché used for people like Robyn who work selflessly for good causes is "tireless." Well, Robyn was not tireless. She was often tired and sick. Lack of oxygen slowed her down. But it did not stop her. After each bout with asthma or bronchitis, she would assure me that, this time, she had things under control. She shared her optimistic and radiant vision of a just and sustainable society with anyone who would listen.

The level of member participation in either growing or distributing the food varies tremendously from farm to farm. At one extreme are CSAs that require all sharers to do some work as part of their share payment. At the other are what have come to be known as "subscription" CSAs, where the farm crew does all the work and members simply receive a box or bag of produce each week. Most CSAs range somewhere in between, with members volunteering for special work days on the farm, helping with distribution, or defraying part of their payment with "working" shares.

Just before she died, Robyn appealed to Community Supported Agriculture groups all over the country to send her copies of their brochures, letters, and newsletters. I spent three full days reading through the material from almost every state and several Canadian provinces. Five years ago, when I started helping Robyn with this book, we thought we would be able to interview someone from every CSA. With each growing season, the projects have proliferated to the point where no one person can do justice to them all.

Reading through the brochures was an uplifting experience—so many eloquent and thoughtful statements about the importance of growing food in harmony with nature and preserving the land for future generations. I covered the floor of my room with them, sorted first by geography, then by size of farm, then by type of project (work required or not, with core group or without, mechanized or worked by hand or horse, weekly packet delivered or picked up, subscription program or community farm), then by years of operation. So many possible categories. So many creative variations on the unifying theme of fresh, local food "with the farmer's face on it," as the Japanese organic farmers would say. So much hope, tempered by a sober awareness of the enormous scale, productivity, and destructiveness of the dominant food system. So much buoyant optimism, despite the overwhelming odds and countless hardships and obstacles.

From the Genesis Farm Community Supported Garden in New Jersey we hear: "All of us are increasingly aware that the present state of farming in the United States and around the world is in serious difficulty. Agribusiness, with its chemical-intensive, industrialized methods of crop production is replacing agriculture, with its small human-scale, and diversified character and its commitment to place and community.

"Many people feel a keen sense of separation from nature and the soil resulting from our urban and suburban settlements. From childhood throughout our adult lives, we feel a deep hunger for a return to the soul of nature. The teeming life, the extraordinary beauty, and the sense of wholeness that a farm provides is a precious gift to nourish this spiritual hunger. The garden offers a way to experience the mystery of seeds and soil and to reconnect in an endless variety of possibilities for creating friendship, community, and the strong connections that historically tied farmers and communities into harmonious relationship with the Earth and each other."

From Goranson Farm in Maine we hear:
"Community Supported Agriculture provides a mutually beneficial arrangement between farmer and community. In exchange for financial support in the spring from the 'shareholders,' the farmers commit to provide healthy, locally grown food through the growing season. The goal of CSA is to reconnect people with the land that sustains them. Shareholders know how and where the food they eat is grown, and they can learn to understand the complexities of providing this food.

"Joining a CSA leads to a greater awareness of our interdependence on one another and the land. It helps ensure the survival of rural life. Our farm has the ability to grow and provide food for a number of families while pursuing proper land stewardship. It is through the cooperation between farm and community that a sustainable local food supply will become a reality."

CLEMENS KALISCHER

CSA members assembling their share basket at Caretaker Farm in Massachusetts.

From Harmony Valley Farm in Wisconsin comes this:
"Again this year, Harmony Valley Farm is making the same, high-quality produce available at a less-than-retail cost to households who wish to participate as CSA members. CSA creates a direct relationship between you and the farm. You will receive a weekly box of our fresh, in-season produce delivered to a convenient location in your neighborhood. You'll know your produce dollar goes directly to the people who plant, tend, and harvest your food. You will be supporting organic growing methods that protect soil resources and water quality and assure you of the most healthful, nutritious produce possible. For everyone, including those of us at the farm, CSA is an opportunity to connect in a meaningful way to others who care about the food we eat."

From Winter Green Community Farm in Oregon this echoes:

"To further our vision of becoming 'a farm in balance with the earth, humanity, and ourselves,' we want to build the link between ourselves and the people who eat the food we produce and to help reestablish the role of agriculture in the community. We seek to provide an environment where families can strengthen their connection to the earth that sustains them. We believe that it makes more sense to grow food for the local community than for distant markets. By joining, you receive fresh, locally grown vegetables and have a direct connection to the farm and the people growing food for you. If you like, you can come and get your hands in the soil, walk the farm, or attend our farm events. In many ways, the most important reason for any of us to be involved with the Community Farm is that it is an affirmation of the kind of world we want to live in. It is a positive choice for the future."

And even from distant Palmer, Alaska, from Arctic Organics we hear:

"Community Supported Agriculture is an answer to what is becoming a worldwide concern: the production and distribution of high-quality, carefully produced food. By joining a CSA program you are supporting your food producers directly by avoiding the middlepeople (i.e., the distributor and retailer). You are purchasing organically and locally grown produce, thereby avoiding the high environmental and health costs and questionable merits of agricultural chemicals *and* the fossil fuels and other resources necessary for shipping it long distances. In exchange, you will receive high-quality, nutrient-rich, flavorful produce on a weekly basis, freshly harvested the same day it's delivered."

CSA members all across the country appreciate the fresh vegetables, but understand that their involvement with a farm means much, much more. Their reflections reveal the potential power of a food system that reconnects people with the land.

From Moore Ranch CSA in California, we hear:

"We're getting good, clean vegetables. And you're not just buying the produce. You're supporting Steve's livelihood. The whole idea is renewing of the Earth."

Harmony Valley members in Wisconsin say:

"We feel better, healthier, and we know where our food comes from."

"My children view Linda and Richard as their farmer. They understand how and where we get what we eat."

At Seven Oaks Farm in Vermont, member Martha Rosenthal writes:

"Janii and Willy . . . are artists in farming and I feel privileged to eat their 'canvas'."

Willow Pond Farm
Sabattus, Maine

In Maine, Willow Pond Farm member Marc Jalbert comments:

"It's very comforting to know where your food is grown. Unlike a symbolic investment in stocks and bonds, we can actually go out and *see* the results of our investment. It's very tangible. It gives me a personal sense of security."

At Happy Heart Farm in Colorado, member Susanne Edminster exclaims:

"The farm itself is miraculous. How you grow all you do on land that seems miniscule in proportion to the bounty is awesome. And there's something about popping a radish or a leaf of lettuce into my mouth, knowing that I don't have to worry about pesticides, which is pure heaven. Sorting and processing the vegetables when I got home was not the dull task I expected. There was a harmonious feeling about it, knowing who grew them, and the love put into the effort. I enthusiastically renew my membership."

IN IOWA, A CSA MEMBER declares: "We became members to support local food production. Our family received high value for our membership in all areas. Economically and healthwise, we felt that the CSA membership improved our eating habits and lowered our grocery bills."

A MEMBER OF THE GENESEE VALLEY ORGANIC CSA in New York State, Josh Tenenbaum, wrote this letter to my partner David and me: "Growing food with the care and loving that you do can never be captured in organic or any other kind of certification. Such things are written only in our hearts. You are both visionaries, and by showing me a brief glimpse of your vision have bestowed on me a wondrous gift. It is a vision of how the world can be—not a dream world, but *this* world. A world of cooperation, of soil teeming with life, of noticing the natural cycles and epicycles, day alternating with night, but also planting, germination, growth, harvest, decay, and then planting again. Thank you for this gift of vision. I ask myself, how do my actions every day create or hinder such a world?"

At the time of this writing, I do not know exactly how many CSA projects there are in the United States or North America. The Biodynamic Association maintains a database of CSAs that lists over six hundred. In *Farms of Tomorrow Revisited*, author Steven McFadden estimated that in 1997 there were over one thousand CSAs feeding one hundred thousand households. When I asked him where he got that number, he admitted he had extrapolated from past figures and rates of growth. I do know that CSAs vary in size from three shares to over eight hundred. They can be found as far north as Palmer, Alaska (and even farther north in Canada), and as far south as Gainesville, Florida, or San Diego, California. The densest clusters are in the Northeast, around the Twin Cities and Madison in the Upper Midwest, and in the Bay Area of California. The number of CSAs is increasing quickly in states such as Iowa, where food activists have teamed up with the Cooperative Extension Service and the universities to provide technical assistance. Most CSAs are either organic or Biodynamic in their method of production. A few are in transition to organic or to a lower use of chemicals. The CSA concept spread from farmer to farmer and from consumer to consumer through the organic and Biodynamic networks, and only recently have a few organizations and Extension agents reached out to conventional farmers. Nothing about the structure of a CSA dictates that the food be organic, but most of the consumers who are willing to become members do not want potentially toxic synthetic chemicals used on their fresh, local produce.

The very first CSAs in this country, Indian Line Farm in Massachusetts and the Temple-Wilton Community Farm in New Hampshire, both initiated in 1986, established the model of the "community farm," which dedicates its entire production to the members, or sharers. Indian Line divided its produce so that every sharer received an equal share or half-share. Temple-Wilton allowed sharers to take what they needed regardless of how much they paid. Only about a quarter of the farms that have adopted the CSA concept have emulated this model. Out of the thirty CSA farms in Vermont, only one produces exclusively for sharers, while the others continue to sell to a variety of markets. The level of member participation in either growing or distributing the food varies tremendously from farm to farm. At one extreme are CSAs like the Genesee Valley Organic in New York, for which I am one of the farmers, and Silver Creek Farm in Ohio, which require all sharers to do some work as part of their share payment. At the other end are what have come to be known as "subscription" CSAs, where the farm crew does all of the work and members simply receive a box or bag of produce each week. Most CSAs range somewhere in between, with members volunteering for special work days on the farm, helping with distribution, or defraying part of their payment with "working" shares.

Jered Lawson, who initiated and then staffed CSA West from 1994 to 1996, expected that CSAs would follow the community farm model, and was at first disillusioned when large organic California farms added subscription shares to their other marketing efforts. He feared that this model would limit member involvement and the sense of connection to the farms. The larger number of members in a subscription CSA certainly reduces the intimacy of personal contact between members and farmers. Compared to smaller farms in parts of the country with more difficult weather conditions, California CSAs have also placed much less emphasis on members sharing the farmers' risks. Dru Rivers of Full Belly Farm told me that she is nervous about pushing the risk-sharing aspect of CSAs because organic food is so readily available in California and there is so much competition. Full Belly Farm confronted this issue once, when a freak snowstorm kept them from picking and delivering one week's shares. Only a few of the members objected to sharing the risk by skipping that week's box. Farms in other parts of the country reduce the risk sharing by purchasing crops from other farms to fill out their shares.

As these larger CSAs have evolved, however, Jered has observed that, far from being a marketing add-on, the community supported component is making an important contribution to the long-term viability of the farms. Somewhat grudgingly, Jered admits that the farmers have adapted CSA effectively to their particular situation:

> Even in a hybrid CSA, or a CSA that may not have the whole farm budget as its centerpiece, there's still the fundamental desire to see the farm

it's all here

we set the seeds, speak
to the sky
nurture the plants, drink
the rain, give back
to the soil, curse
the cold, dance
to the sun, sing
with the wind, weep
at the passing, dream
with the moon. we open
our hands and accept another
season of hope fulfilled
or not,
balancing burdens with blessings,
rocks and eagle feathers,
carrying the harvest home.
listen.
the birds are singing the earth
awake. the spiralling cosmos
is bursting open seeds climbing
to the light. there's a crackle
of joy in our hearts, ignited
by the sun—a flower filled
with flame. listen. the plants will tell you
of sending roots deep to survive
the dry times.
the seasons will show you how nothing
is ever really gone but keeps
turning out and over
again and again and again,
just as the ancestors
smile down from the clouds
onto the faces of children
yet unborn shining up
from rain spattered stones on the path
we walk.
listen.
it's all here.

—Sherrie Mickel, 1995

> survive. The consumer members of the farm do come to understand that they are giving economic contributions towards that survival, and the farmers themselves help set the terms of that contribution. Therefore, the consumers can say "we feel comfortable that we are meeting the farm's assessment of what it needs to survive," even though it is different from the original CSA concept. (McFadden and Groh, p. 83)

Through their teachings and example, Robyn Van En and Trauger Groh shaped the "original CSA concept." Robyn tells the story of founding Indian Line Farm CSA in the Introduction to this book. Trauger Groh has been one of the team of three farmers at Temple-Wilton Community Farm through its entire history as a CSA. He is also one of the authors of *Farms of Tomorrow* (1990) and *Farms of Tomorrow Revisited* (1997), which have inspired many CSA efforts. (You will find these sources and others cited throughout this book in the Bibliography.) In these two books, Trauger and his coauthor Steven McFadden profile a number of CSA farms, all of them adherents of Biodynamics, an approach to farming based on the writings of Rudolf Steiner and his followers. I recommend that anyone interested in CSA read *Farms of Tomorrow Revisited.* Biodynamic principles and the farms that apply them have a great deal to teach the rest of us in agriculture. But you do not have to adhere to Steiner's precepts to practice community supported agriculture.

Robyn and I agreed that the purpose of this book should be to help spread community supported agriculture by collecting as many of the best examples we could find regardless of ideology, religion, farming methods, or sexual preference. We did not want to craft a tight definition or try to establish the criteria for identifying "the true CSA farm." Rather, we hoped to honor the diversity of this young, but quickly spreading movement. Like the farms that host them, CSAs come in many scales and sizes. The people growing the food have different ideas about how to do it, whether efficiency is a value, how much they need or want to earn, and how many helpers they organize to get the work done. I like the way Steve Gilman talks about CSA diversity in "Our Stories": "Like grapes or garlic, CSA takes on the flavor, bouquet, and integrity of where it grows, becoming appropriately adapted to each unique situation." While participants agree that CSA means a connection between a specific group of eaters and a specific piece of land and the people who farm it, healthy debates roar on about how to understand the concepts of "community," "support," and even "agriculture." You will hear passionate snippets of those debates in this book.

As an organic farmer active in farm politics, I am used to being in rooms or fields full of opinionated and articulate people. Sometimes we all talk at once. Lately, we have been learning to set an agenda with amounts of time for each item, so that we can speak one at a time and practice the active listening skills we are also struggling to acquire. I propose to do something like that with this book—to let many different versions of CSA speak for themselves. In Robyn's place, I will give as much as I have been able to salvage of her version and chime in with mine as well.

CSA AND THE GLOBAL SUPERMARKET 2

America of the future will be all malls connected by interstates.
All because your parents no longer can their own tomatoes.

—GARRISON KEILLOR
NATIONAL PUBLIC RADIO, MARCH 28, 1998

I GREW UP IN CROTON-ON-HUDSON, New York, a bedroom suburb of New York City, raised by parents who were deeply committed to the struggle for world peace and economic justice. They were city people through and through, as were my grandparents. No one even gardened, but at the dinner table we had long, disturbing conversations about world politics, hunger, malnutrition, and inequality. A few miles up the Hudson, the government stored millions of tons of surplus grain in cargo ships anchored along the shore. Playing hooky from school, I sometimes drove along the scenic river and puzzled over those hulking grey vessels. If all that surplus was such a big expense and there were so many starving people, I reasoned in my innocence, why couldn't they just sail those boats to wherever food was scarce and feed the hungry? The irrationality of the food system seemed unfathomable.

After forty years and a lot of travel, reading, and thinking, I see some pieces of the picture starting to fall into place. I do not make any claim to be an expert analyst of the world's food and agriculture systems, but after almost two decades of living and working as an organic farmer this is how things look to me.

Food is basic to human existence. "We are what we eat." We can go a day or two without eating, even a week; most people worldwide, however, prefer at least one daily meal. "Our bodies," as Bill Duesing, an organic educator in Connecticut, explained on the "Living on Earth" radio program, "run on the solar energy collected by plants, which are nourished by the nutrients they gather from water, air, and soil." Farmers are "managers of solar conversion," to borrow a term from Holistic Resource Management. In cooperation with the forces of nature, farmers and gardeners actually create wealth. For most of human history, human beings have not taken their food supply for granted and have regarded the creation of food as a sacred act, surrounding it with rituals of blessing and expressions of gratitude to the Earth or the gods or God. Growing food is the most basic use of the natural resources of the Earth, and through food production, we make our own working landscapes. How each society or nation produces and distributes food in large measure determines its identity.

Our farm and food industry leaders tell us that the U.S. food system is the safest, cheapest, and best in the world. Not only do American farmers feed the growing population of this country, they keep the rest of the world from starving as well. With barriers to trade cleared away by the international GATT and NAFTA agreements, U.S. farms, we are told, will be able to outcompete all others. The $57 billion worth of agricultural exports in 1997 could rise to $75 billion in the near future as the standard of living rises in China, boosting purchases of U.S. grains and meat products. Twenty-four million acres were planted with genetically modified seed in 1997, and that is only the beginning. A team of researchers at the Department of Energy's Pacific Northwest National Laboratory predicts that *agrogenetics*, genetic engineering combined with plant manipulation, will reduce agricultural impacts on the environment. The library of genetic information is doubling every twelve to twenty-four months: Monsanto, the "life sciences" company, can create ten thousand new genetic combinations a year, according to Barnaby Feder in the *New York Times*. Genetically engineered crops will resist pests, requiring less pesticides, and will assimilate nutrients more efficiently, reducing the need for fertilizers. By using custom-designed, genetically engineered seed varieties, global positioning computers, and precision farming technologies, as few as fifty thousand farmers will supply 75 percent of the country's agricultural production from 50 percent of the farmland. Ray Goldberg writes that these farmers will work as "farm technology managers" under contract to the vertically integrated distributors and processors. Armed with scanning feedback devices providing them with unparalleled amounts of intelligence on consumer finances and purchasing patterns, distributors will become the market coordinators of the domestic and global food systems.

How does this upbeat presentation fit with some of the other realities of our food supply? While the stores are well stocked with food and there are no obvious shortages, the supply of food on hand in Northeastern cities would only last thirteen days should some emergency occur. A 1997 USDA report revealed that almost twelve million U.S. households may be "food insecure." A Cornell University study headed by Katherine Alaimo, based on a survey conducted between 1988 and 1994, confirmed that ten million Americans do not get enough to eat, including four million children. In the U.S., the number of food banks, community warehouses distributing salvaged and donated food to emergency food providers, rose from 75 in 1980 to 225 in 1990, and the number of pounds of food distributed rose from 25 million in 1980 to 811.3 million in 1995. Second Harvest reports that almost twenty-six million people rely on the emergency feeding programs their network serves. Yet Congress has responded by reducing funding for food stamps! Lester Brown of World Watch Institute reports that, according to the World Bank, 1.3 billion people worldwide have less than $1 a day to spend on food, a mind-numbing level of human misery.

Within the memory of people still alive, fruits and vegetables grew on the outskirts of most major cities. Market farms across the Hudson River in New Jersey supplied the vegetable

Growing food is the most basic use of the natural resources of the Earth, and through food production, we make our own working landscapes. How each society or nation produces and distributes food in large measure determines its identity.

needs of New York City dwellers. Apple orchards blossomed in the Bronx. Since that time, urban and then suburban sprawl have paved over the cropland and planted houses and highways where cabbages used to grow. Food production has shifted to where it is most "efficient," where bigger machines can maneuver over larger, flatter fields, where chemicals and technology reduce the need for horse and human power, and where crews of poorly paid migrant workers do their jobs and then move on. The global food system plays economic hardball. A vicious speed-up has been going on in the countryside. Where a farmer could support a family with a herd of twenty cows in 1950, today two hundred are needed. Where once a 160-acre "section" produced enough grain for a family's livelihood, today 1,600 acres with a much greater yield per acre is just barely enough to keep a farm in business.

Increasingly large and specialized farms produce basic commodities, yet rural areas no longer feed themselves. The Field to Family Community Food Project states:

> Iowa is a "textbook example" of the effects of an expansive industrial food system. The state's agricultural production, focused on grain and livestock, is highly specialized for export purposes, and because there are few food processing industries based in the state, almost all of the food consumed by the state's 2.8 million citizens (including that derived from the basic commodities produced in the state) is imported. The state depends on such imports for essentially all of its vegetables and fruits.

Until the 20th century, the United States was largely an agrarian society. As late as 1910, one-third of the population, some thirty-two million people, lived on farms. The Dust Bowl, the Depression, and the World War II drove nine million people off the land. But those major upheavals pale in significance compared to the decimation of the farming population that followed the restructuring of farm price supports in the 1950s. By 1993, less than 2 percent of the population was left on only 2.2 million farms, so few that the U.S. Census Bureau announced it would stop counting them. The number of farms fell again to 1.92 million in the Agricultural census of 1997 (*American Farmland*, p. 4). Black farmers have been squeezed off the land even faster than white farmers: In 1920, one out of every seven farmers was black; in 1982, the National Land Loss Fund indicated that black farmers counted for only one out of sixty seven and operated only 1 percent of the farms. Between 1987 and 1992, Vermont lost seventy-three acres of farmland a day. New York State has been losing farms at the rate of twenty a week and farmland at the rate of one hundred thousand acres a year for twenty years. The country lost 8 percent of its dairy farms in the years 1996 to 1997. According to the National Agricultural Lands Study completed in 1981, the U.S. was losing one million acres of prime cropland every year, or four square miles a day. Julia Freedgood of American Farmland Trust says that this rate of loss of three thousand acres a day is continuing in the 1990s.

Farmers have a saying: Farmers sell wholesale and buy retail. The terms of this deal get worse and worse. The index of prices farmers pay for seed, equipment, and other necessities has risen 23 percent since 1950, while the prices paid to farmers at the farm gate fell 60 percent. The value of the basic commodities produced by farms is sinking: Between 1978 and 1988, the wholesale price of milk fell by 11 percent, potatoes by 9 percent, fresh vegetables by 23 percent, and red meat by 37 percent. In 1981, dairy farmers received $13.76 per hundredweight (the national average), while consumers paid $1.86 per gallon of milk. In 1997, farmers were paid $12.70 per hundredweight, but the consumer price was up to $2.76 per gallon (USDA Economic Research Statistical Bulletin #849, December 1992).

The complex and confusing federal system of loans, set-asides, deficiency payments, and the like has not resulted in prices that cover the cost of the reproduction of the farms. Let me give an example: In 1993, the *annual* cash expenses for growing an acre of corn, as calculated by the Economic Research Service of USDA, amounted to $177.89. The total gross value of selling that corn was $227.36, for an apparent profit of $49.47 an acre. However, when the government economists added in the *full* "ownership costs," expenses that must be covered to keep a farm economically viable for the long run—such items as capital replacement, operating capital, land, and unpaid family labor, the bottom line was *minus* $59.74, and this did not include money put aside for retirement. An acre of oats came in at minus $51.18, wheat minus $52.87, and milk, per hundredweight, at minus $2.02.

According to the National Agricultural Lands Study completed in 1981, the U.S. was losing one million acres of prime cropland every year, or four square miles a day. Julia Freedgood of American Farmland Trust says that this rate of loss of three thousand acres a day is continuing in the 1990s.

Small surprise that farms all over the country are going out of business. In a little booklet published in 1979, "The Loss of Our Family Farms," Mark Ritchie asks whether this is the inevitable course of history or the result of conscious policies. He concluded that the loss of so many farms is not the unfortunate result of policies that failed, but rather the result of a concerted and unrelenting drive by agribusiness, government, banking, and university forces to restructure agriculture by reducing farm price supports, manipulating the tax structure, and conducting research and development in support of large-scale agricultural enterprises. Ritchie identifies the men who made these policies as representatives of the largest corporations, banks, and universities, who saw their work, in their own words, as contributing "to the preservation and strengthening of our free society."

Middle-sized farms with gross sales ranging from $10,000 to $99,999 a year are disappearing the fastest. In 1997, farms selling less than $9,999 increased by 1 percent and farms selling over $100,000 increased by 3 percent, while the middle group fell by 4 percent. In *Family Farming: A New Economic Vision*, Marty Strange presented convincing evidence that farms in the middle category, particularly the $40,000 to $250,000 range, make better use of their resources and are more likely to practice careful land stewardship than the largest farms. For every dollar that the family-run, middle-sized farms spend, they produce more income. Production expenses on the largest farms averaged 85 percent of gross sales, while the middle farms averaged only 72 percent. On organic CSA farms such as Rose Valley, production expenses are even lower, in the range of 40 to 50 percent of gross sales. Like Ritchie, Strange concluded that the obstacles to the survival of these farms come from public policy, not from poor farm management or lack of efficiency. Since the 1950s, public policy has been pushing farms to "get bigger or get out."

In 1979, the USDA issued a report entitled *A Time to Choose*, which warned that "unless present policy and programs are changed so that they counter instead of reinforce or accelerate the trends toward ever-larger farming operations, the result will be a few large farms controlling food production in only a few years." Eighteen years and over three hundred thousand lost farms later, the USDA convened a National Commission on Small Farms to examine the condition of farming and its place in the food system. The commission held a series of hearings around the country where, on very short notice, hundreds of farmers and farmer advocates testified. In January 1998, the commission released its report, *A Time to Act*. This report contains 146 recommendations for policies that would protect small farmers' access to fair markets and redirect existing federal programs, which are currently skewed to serve the interests of large agri-businesses. Written by Barbara Meister, an active member of the National Campaign for Sustainable Agriculture, *A Time to Act* contains stirring language about the vital role of small farms as the embodiment of the ideals of Thomas Jefferson, and it is perfect for waving in the faces of legislators when grassroots activists lobby for the changes it suggests. Hopefully, nineteen years hence, it will not be gathering dust next to old copies of *A Time to Choose.*

Ocean Sky Farm
Bainbridge Island, Washington

As overall farm numbers shrink, the remaining farms have industrialized, attempting to dominate the biological process of growing food with chemical and mechanical technologies, and achieving spectacular results in terms of production per acre. With chemical fertilizers, synthetic herbicides and insecticides, hybrid varieties, irrigation, and ever larger machinery, yields of basic grains—corn, wheat, and rice—have doubled and tripled.

The environmental costs of this bounty are staggering. When the Environmental Protection Agency (EPA) tested rivers, lakes, and wetlands around the country, they found that barely half could support all uses: clean water for drinking, swimming, and recreation; fish and shellfish that are safe to eat; habitat for healthy aquatic wildlife. Agriculture accounted for 72 percent of the pollution of rivers and streams with silt, runoff from fields, and excess nutrients, such as phosphorus. Forty-six different pesticides and nitrates from nitrogen fertilizers have been found in the groundwater of twenty-five states, with the largest residues in big agricultural states, such as California and Iowa. Evidence is mounting in support of the thesis Theo Colburn put forth in *Our Stolen Future,* that chemicals used as pesticides act as endocrine disruptors, upsetting the normal hormonal development of frogs, seagulls, polar bears, and human beings. As *Consumer Reports* puts it in a study of pesticide residues in food, "no one really knows what a lifetime of consuming the tiny quantities of pesticides found on foods might to do a person." While conclusive proof of this connection is still lacking, the pervasiveness of these chemicals in the environment and in our bodies should be cause for enough alarm to jar us into action.

In a fine essay on CSA in the collection *Rooted in the Land,* Jack Kittredge sums up the loss of soil to erosion:

> Much of the incredible productivity of North American industrial agriculture has been based on using up two irreplaceable capital assets: topsoil and petroleum. When our ancestors settled this continent, they benefitted from the largesse of thousands of years of natural soil creation, virtually undiminished by agriculture. The careless practices of our modern world have seen up to half of that soil washed or blown away, and each year every acre loses an average of 7.7 tons more. Only when the last of this prehistoric legacy is washed to sea and we are on a "level playing field" with the other continents will we appreciate the magnitude of our folly. (p. 26)

Jim Hightower has the clearest grasp of the logic behind the industrial farming approach: "If brute force isn't working, you're probably not using enough of it." Insects, diseases, and weeds were reducing crops worldwide by 34.9 percent in 1965. Pesticide applications in the U.S. totalled 540 million pounds of active ingredients. By 1990, the National Center for Food and Agriculture Policy reported that losses to pests rose to 42.1 percent although U.S. farmers poured on 886 million pounds of pesticides. Pesticide company profits continued to climb in 1997. Novartis, the largest pesticide company in the world, sold $4.2 billion worth of agrochemicals, up 21 percent from 1996, according to the Pesticide Action Network of North America. In the *State of the World* for 1998, Lester Brown argues that the world has surpassed the environmental limits to continued increase in agricultural yields: Soil erosion plus declining underground water supplies plus climate change due to global warming overbalance any further benefit from additional fertilizers, pesticides, or improved varieties. This is the reality of a too-full world.

The biggest farms may be getting bigger, but the farming sector of the food system is losing control to the increasingly consolidated multinational corporations such as Novartis. Many of the once self-employed farmers become employees at larger farms or in the farm inputs, processing, and marketing sectors, which are returning 18 percent on investment, and grabbing ever-larger portions of the consumer food dollar from the farms. Stewart N. Smith, former Commissioner of Agriculture for Maine and an agricultural economist, has traced the downward trajectory of farming:

> The food and agriculture system has changed remarkably through this century under the regime of industrial agriculture, especially in shifting economic activities from the farm to the non-farm components of the system. Farmers contributed 41 percent of the system activity (and got 41 percent of the returns) in 1910, but only 9 percent in 1990. On the other hand, input suppliers increased their share from 15 percent to 24 percent, and marketers from 44 to 67 percent.

According to Smith's calculations, if current trends continue, farming as such will disappear completely in the year 2020.

The passage of GATT and NAFTA remove what few protections were left for U.S. agricultural producers. Under GATT regulations, U.S. agricultural exports have risen 5 percent, but imports have gone up 32 percent. In 1997, this country imported $36.2 billion worth of food products, and that wasn't all coffee, chocolate, and bananas. It included many crops we can produce right here. According to a USDA study released in 1997, the economic impact of NAFTA on the balance of agricultural trade between the U.S. and its two neighbors has been a negative $100 million (cited by Alan Guebert in *Agri News*). Mexican tomatoes are underselling Florida tomatoes. U.S. corn growers, however, favor NAFTA because it opened Mexican markets to them and even helped raise the price of corn 8 cents a bushel. The cost has fallen on Mexico, where Steve Suppan and Karen Lehman report that six hundred thousand to one million small corn farmers, who could not compete with the lower price of the imported corn, have been uprooted from the countryside and forced into the army of unemployed in the towns.

When international trade heats up, the greatest benefits go to the corporations that control the markets. Fewer than five companies control 90 percent of the export market for corn, wheat, coffee, tea, pineapple, cotton, tobacco, jute, and forest products, according to William Heffernan. Those same big traders—Cargill, Continental Grain, Bunge, Luis Dreyfus, Andre and Co., and Mitsui/Cook—also control storage, transport, and food processing. Incidentally, Daniel Amstutz, a former Cargill executive, drafted the U.S. agricultural proposal for GATT under President Reagan.

The consolidation of control of the food system inside the U.S. is also increasing steadily. Tom Lyson has calculated that 10 cents of every food dollar spent in this country goes to Philip Morris, a conglomerate of nine tobacco brands, and the owner of Miller Brewing, 7-Up, Post Cereals, Maxwell House Coffee, Sanka, Jell-O, Oscar Mayer, Log Cabin Syrup, and more, with total sales in 1995 of $36 billion. As Karen Lehman and Al Krebs put it:

> Between January 1 and January 31, 1995, while most Americans were still figuring out how to break their New Year's resolutions, Philip Morris merged Kraft and General Foods into Kraft Foods; Ralston Purina sold Continental Baking Company to Interstate Bakeries Corporation, the

family business

both drizzled with grey and not so slim around the middle
anymore, a woman and a dog endure the heat
side by side in a meadow shimmering life. they have
shared many miles of the Good Red road, 4 feet 2 feet
plod and dance, trot and stumble, lope and scramble
right on down to this afternoon hunkered over under
the sun so hot. the dog's head is nestled in sweet
grass at the streambank, the woman's is bent over
hairy weeds surrounding baby lettuces, and deer flies
whiz between them like agents of a curse. "sissies,"
hiss the hairy weeds; they sneer, "we will smother
you." and the woman's hands tan as the soil and just
as lined pull them out one by one and tuck them in
beside the babies.
"why is it the bad so often seem so strong
while the good get by on grace?" she asks the dog who
smiles and wags but does not raise her head. so the
woman calls out clear as crows discussing family business
across the pasture at first light, "what I mean is, how
can something that shines so true and mighty be so fragile
when shadows pass and blot out the brightness so easily
before passing away again?"
and the dog digs thoughtfully into her ear
limned with grey, cloudy brown eyes focused on something
near the horizon that the woman lifts her gaze to see
but doesn't find. she speaks once more but quiet now
as the stream murmuring to itself. "what I do understand is
you, old dog. you and the sun and these damned flies, these
hairy weeds and baby lettuces. I know what to pull and
what to save, and where to put them all. and just who will
sit out here beside me in this heat until the work is done."

—Sherrie Mickel, 1997

nation's largest bread maker; Perdue Farms, the nation's fourth-largest poultry producer, acquired Showell Farms, the nation's tenth-largest poultry producer; and Grand Metropolitan proposed to acquire Pet, Inc. The brand names are all that are left of the small companies that became huge conglomerates through mergers and acquisitions.

The four major packers control 86 percent of the beef slaughter, the greatest concentration in U.S. history. What does this mean for consumers and small farmers? When there are only one or two buyers, farmers have to take what they can get if they want to sell their crops. If they don't cooperate with the big packers, farmers can find themselves without any buyers at all. Between 1979 and 1997, the producer share of retail beef sales dropped from 64 percent to 49 percent, as the price farmers received for slaughter steer fell 50 percent. Consumers ended up paying less, but only by 15 percent, while the packing companies enjoyed unprecedented profits (see the *Red Meat Yearbook* and *Food, Farm and Consumer Forum*). Anti-trust legislation, the Sherman Act of 1890, requires that when as few as four companies gain control of 60 percent of any sector, the government must take action. Passed in response to public anger over concentration, the act assumes that dominance of an industry or market by a few firms will damage the public by raising prices, reducing quality, and slowing technological advance. Obviously, the government has chosen not to enforce this set of laws very often.

We are living through the culmination of the era of the transnationals, the megacorporations such as Philip Morris, Cargill, Archer Daniels Midland ("Supermarket to the World"), and ConAgra. They have no allegiance to any particular nation or group of nations, and they are manipulating the rules of the World Trade Organization (WTO), established by the GATT, to increase their power to intervene in the economy of any country as they see fit. The power of the WTO was demonstrated when it struck down as a barrier to trade the European Union's ban on U.S. hormone-treated beef. President Clinton wanted the fast-track authority to expand NAFTA, but also to negotiate the Multilateral Agreement on Investment (MAI). The MAI, "GATT on steroids," is the final link in the choke collar the multinationals have been forging behind closed doors. If implemented, the MAI would give corporations the freedom to do business in any country without the control of the national government. If a government objects that a multinational is violating local environmental or labor laws, the multinational will be able to take that government to court.

From the consumer's point of view, the source of food lies hidden behind an almost impenetrable wall of plastic and petroleum. Few stores bother to label food with its point of origin. The 7 to 10 percent value of the raw food in processed products is buried by the other 90 percent—the chopping, blending, cooking, extruding, packaging, distributing, and advertising. Gary Argiropoulos, vegetable and floral manager for Hannaford Brothers in Portland, Maine, told me that "fresh" and "local" in the language of supermarket produce buyers means accessible within twenty-four hours by air freight. I once heard Michael Osterholm, the Minnesota State Epidemiologist, talking about food safety on National Public Radio. He attributed the increase in foodborne illnesses, at least in part, to eating out-of-season produce shipped from countries with low health standards; to so many of the food workers being low-paid, uneducated, and lacking in proper health care; and to the rise of drug resistance in microorganisms such as camphylobacter in antibiotic-fed chickens. Judith Hoffman has summed up well the widening gulf that the global supermarket leaves between farmer and consumer:

> The *inter-* of our interdependence is gone and only the dependence remains. This kind of dependence is the essence of insecurity because it permits exploitation. By the time the consumer's dollar has made it to the farmer, there are mere pennies left of it. By the time the farmer's food has made it to the consumer, it has been bled dry of flavor and passed through a gauntlet of dismayingly mysterious chemistry. It seems all that farmers and consumers have left in common as the century turns is anxiety.

THE MOVEMENT FOR A SUSTAINABLE FOOD SYSTEM

Every aspect of our lives is, in a sense, a vote for the kind of world we want to live in.

—Frances Moore Lappé

Deciding to join a movement for a sustainable, regional food system is like jumping into a swiftly flowing river of icy water and swimming against the current. Yet all over the country, thousands of people with their eyes open are doing just this. At first a trickle of isolated farmers resisted the pressures to adopt chemicals and specialize their farming on single crops. Lone consumers, hunting for food that was unadulterated and natural, joined together to form food coops. In the 1970s, small grassroots organizations formed the Natural Organic Farmers

Association in the Northeast, inspired by J. I. Rodale and Sir Albert Howard; the Center for Rural Affairs in Walthill, Nebraska, founded to stop the decline of the family farm; the Federation of Southern Coops, emerging from the civil rights movement to keep black farmers on their land; and many, many others.

Over the past decade, this scattering of local organizations has swelled into a significant social movement with a national network and effective policy wing. In 1995, the National Campaign for Sustainable Agriculture coordinated the efforts of over five hundred member organizations and farms in lobbying for policy proposals for the Farm Bill. In 1998, over 280,000 people sent in comments when the USDA proposed nationwide regulations for organic agriculture that would have made "organic" indistinguishable from conventional industrial production. Populist in spirit, with strong feelings for civil rights and social justice, and an underlying spirituality, the movement for sustainable agriculture is not linked with any political party or religious sect. It is firmly grounded in every region in the country, encompassing organic and low-input farmers; organizations concerned with food, farming, farmworkers, community food security, social justice, and hunger; advocates for the humane treatment of animals; and environmental, consumer, and religious groups. Community Supported Agriculture is an important part of this movement.

Bucking the dominant system both literally and intellectually, the people involved in sustainable agriculture have had to struggle to move beyond reductionist science and "best management practices" for improving farming one problem at a time to thinking in terms of interdependent systems. Biodynamic and organic farmers with their holistic approach to the farm as an integrated part of its ecosystem, as well as practitioners of Permaculture design and Holistic Resource Management, have all contributed to the "paradigm shift." Although still referred to as sustainable agriculture, the term has come to mean food production in the context of the farm's physical, social, and economic environment. On their Web site, the Sustainable Agriculture Network offers this broad and interactive definition:

> Sustainable agriculture does not refer to a prescribed set of practices. Instead, it challenges producers to think about the long-term implications of practices and the broad interactions and dynamics of agricultural systems. It also invites consumers to get more involved in agriculture by learning more about and becoming active participants in their food systems. A key goal is to understand agriculture from an ecological perspective—in terms of nutrient and energy dynamics, and interactions among plants, animals, insects, and other organisms in agroeco-systems—then balance it with profit, community, and consumer needs.

From its origins in this country in 1985, the people who initiated CSA have hoped it would provide an antidote to some of the worst aspects of the prevailing food system. Jan Vandertuin, who brought the idea from Switzerland to Robyn Van En at Indian Line Farm, wrote "Vegetables for All." In this essay, Jan describes his frustrations at trying to find work producing quality vegetables:

> *Good-looking* vegetables are often produced with the use of pesticides and herbicides. *Fresh* may mean storage and transportation methods that are energy-intensive—or questionable, such as irradiation. *Organic* produce often means overworked, underpaid agricultural workers. And *reasonably priced* generally means as cheap as possible without any regard for the hidden costs—government subsidies, market manipulation, exploitation of the Third World, and pollution of the environment.

Jan's search for a situation that gave recognition to the value of agricultural work led him from organic farms and whole-grain bakeries in the United States to the producer-consumer associations in Switzerland, which served as the inspiration for CSA in the southern Berkshires of Massachusetts.

Fitting as a manifesto for the CSAs of the future, the very first CSA proposal sets out these ideals for agriculture:

RESPECT FOR

animate Earth
wild plants and animals
domesticated plants and animals
environmental limits
cycles of nature, including the growing season and animal breeding season
food workers' physical, social, spiritual needs

RESPONSIBILITY TO

Use organic/Biodynamic methods and insights, including

- growing own seeds/seedlings (local exchange?)
- composting kitchen wastes for garden
- garden plan with balanced variety
- quality storing/preserving for year-round supply

Be energy-conscious in production and distribution

- minimal machinery
- local distribution network
- human or animal power

Maintain decent working conditions

- decent wages, i.e., $6–7 per hour
- limited hours, i.e., 35–42 hours per week
- fully integrate and pay trainees

Emphasize the therapeutic value of agricultural work (for non-gardeners)

Support community control of the land

- balanced land use plan
- special provisions for agriculture
- eliminate speculation

Create local social/economic forms, based on trust, which

- encourage initiative and self-reliance
- share the risks of agricultural production
- share information
- are human-scale and efficient
- charge according to needs/cost (not market)
- provide locally controlled financial services (currency, banking, insurance, etc.)

Think globally, act locally.

THE WAY IS AS IMPORTANT AS THE GOAL

As we go through the achievements of CSAs in this book, we should look back on this statement from time to time to see how many of these ideals are being realized.

In *Farms of Tomorrow Revisited*, Trauger Groh reacts against the corporate miasma with an appealing utopian vision of a future in which farmland ceases to be privately owned real estate and farmers cease to be businessmen. As Trauger paints them, the "farms of tomorrow" are individualized organisms attuned to cosmic rhythms and based on three principles:

- the highest diversity brings the highest productivity;
- harmony needs animals for a balance of crops for market and cover crops;
- the farm is a closed system, producing and recycling its own fertility.

The formation of community farms "liberates the farmer to work out of his spiritual intentions, not out of money considerations." The members of these community farms, in Trauger's conception, come together around shared ideals to cover all the costs and share all the risks of their farms, and the farms link together in an "associative economy" in which spirituality prevails:

> In an associative economy, we associate with our partners—active farmers among themselves, active farmers with all the member households, farm communities with other farm communities. The prevailing attitude is a striving to learn the real needs of our partners, and the ways we can best meet them. That means we do not make our self-interest the driving force of our economic behavior, but rather we take from the needs of our partners the motivation of our economic actions. We believe and trust that this will lead to the greatest welfare of all involved. (McFadden and Groh, *Farms of Tomorrow Revisited*, p. 35)

In what Trauger sees as a period of diminishing life-force in nature, the farm radiates life-giving energy. To CSAs, he assigns a broad set of tasks: education of the young, the revival of ethics, and the renewal of human health, culture, the economy, and social life.

While I find Trauger's ideas appealing, it is important to me to run a farm that does function successfully as a business. I want my farm to serve as a demonstration to my farming neighbors, many of them very conservative people, that ecological farming is a practical possibility. I realize that the conventional farmers I know consider my organic CSA to be a sort of special case, but at the same time, they recognize it as a creative approach to marketing and admire my ability to get the cooperation of consumers. Though what I am doing is different, it is somehow still within the range of acceptable local farming practices, and that is a great advance over how it was viewed ten years ago.

We may have to drop some of our favorite jargon, even the word "sustainable," and talk, instead, about keeping farms in business for the long term, making sure that everybody gets enough nourishing food to eat, and living in a way that respects the natural limits of the world around us.

Wendell Berry, who lambastes the system of multinational corporations for its institutionalized irresponsibility, is convinced that the old political parties—left, right, and center—have outlived their usefulness. To bring about a sustainable world, Berry suggests, we must understand that the leaders of all of those parties have sold out to the corporations and really amount to only one party—the party of the global economy, a party which is "large, though not populous, immensely powerful, self-aware, purposeful, and tightly organized" ("Conserving Communities," p. 412). Totally unrepresented by that party are the ordinary people, both rural and urban, who share an appreciation of the value of neighborly acts, the urgency of protecting the purity of local land, water, air, and wild creatures, and the need to build cooperative links between farmers and community-minded people in nearby towns and cities.

We don't seem to know it yet, but CSAs and our more conservative rural neighbors belong together in the second party, what Berry calls the party of the local community. So do the urban people who join CSAs, without any thought about the broader implications, simply to get fresh, organically grown food. I hope we will be able to spread the word about community supported agriculture in language that opens doors to people. We may have to drop some of our favorite jargon, even the word "sustainable," and talk, instead, about keeping farms in business for the long term, making sure that everybody gets enough nourishing food to eat, and living in a way that respects the natural limits of the world around us. We need to find that place

CLEMENS KALISCHER

Sharers and their children weeding early season crops at Indian Line Farm.

where the conservatism in organic farming meets the conservatism of small-town Republicans and Democrats.

Sustainable local food production and community supported agriculture are also essential to sustainable development. Herman E. Daly emphatically distinguishes sustainable development from "sustainable growth," which he terms an oxymoron. In his words:

> An economy in sustainable development adapts and improves in knowledge, organization, technical efficiency, and wisdom; it does this without assimilating or accreting an ever greater percentage of the matter-energy of the ecosystem into itself, but rather stops at a scale at which the remaining ecosystem can continue to function and renew itself year after year. ("Sustainable Growth," p. 195)

Those who want to undertake sustainable development would do well to look upon CSAs as places where farming and non-farming people are learning to create local economies that fit Daly's definition, providing for people's needs while regenerating resources. A farmer interviewed by Jack Kittredge summarized the value of CSAs:

> CSA seems like the best way to establish relationships between people who farm and consumers. It brings appreciation of the other's point

> of view to both sides. It helps break the wholesale/grocery store mindset, which dehumanizes farm work and makes the Earth seem like a natural resource to be used rather than cared for. It helps farmers to have a community (since the community of fellow farmers is so small). It helps provide financial stability to farmers by knowing what the year has in store for them and evens out cash flow. I like it personally, agriculturally, and financially. ("CSAs in the Northeast," p. 12)

Sharing the Harvest will take you into the intense world of community supported agriculture. A world of heady ideals and hard physical labor. A world where people from very different backgrounds and perspectives are struggling to learn how to do practical work together to create a new food system based on priceless values:

- an intimate relation with our food and the land on which it is grown;
- a sense of reverence for life;
- cooperation;
- justice;
- appreciation for the beauty of the cultivated landscape; and
- a fitting humility about the place of human beings in the scheme of nature.

— Part II —

Getting Started

Creating a CSA 3

. . . A world that supranational corporations and the governments and educational systems that serve them . . . control . . . will be . . . a postagricultural world. But as we now begin to see, you cannot have a postagricultural world that is not also postdemocratic, postreligious, and postnatural—in other words it will be posthuman, contrary to the best that we have meant by humanity.

—Wendell Berry, "Conserving Communities"

Starting a Community Supported Agriculture project is a little like having a baby—you unleash biological and social forces that may take you in directions you never expected. CSAs have come into being in many different ways: from existing farms; singly or in groups; for part or all of their production; established by institutions, such as land trusts, religious orders, or food banks; or improvised by would-be farmers or groups of consumers. There are many common elements, but each birth is unique. Here are some creation stories.

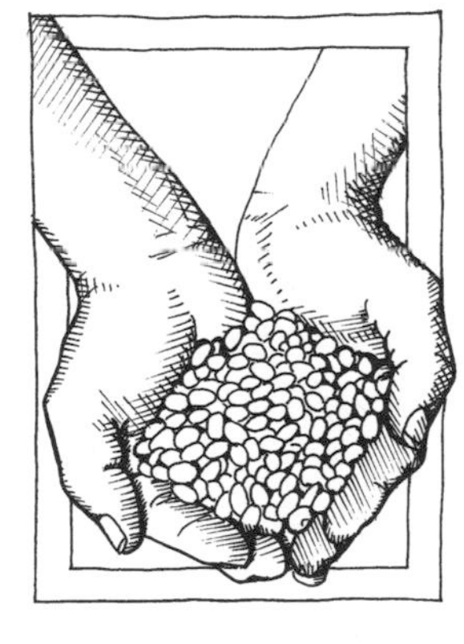

The origins of the Genesee Valley Organic CSA (GVOCSA) reach back to a box of organic vegetables I found on my doorstep in France during the summer of 1977. I had sublet an apartment from friends in the walled town of La Cadière in Provence. They had told me they were Maoists, but they had not told me that their cell was supporting two farmer comrades who delivered a box of vegetables once a week. I was already toying with the idea of market gardening, so curiosity about the box quickly led me to visit the little rented farm. My most vivid memories are of the cement-like soil and baby goats on the roof of our car. Unfortunately, I do not know what became of that French-Maoist CSA.

The idea stayed with me, though, and when I moved to Rose Valley in 1988 and heard about Indian Line Farm, I proposed to my partner David that we try a CSA. During the winter of 1989, David and I had a meeting with Alison Clarke, founder and staffperson of the Peace and Justice Education Center (today the Politics of Food, or POF) in Rochester, New York, and PJEC member Jim Marks, a Xerox engineer. We agreed to the broad outlines of the project and decided to put out a flyer inviting people to an organizational meeting.

By today's standards, the flyer was pretty crude, half-typed with headlines written in by hand, but when posted around the Genesee Valley Food Coop and Ozone Brothers store in Rochester, and mailed to the members of PJEC and the local chapter of the Human Ecology Action League (HEAL), it attracted twenty-four people. The flyer read:

> Do you want fresh-picked, organically grown vegetables? Join in Community Supported Agriculture, a cooperative experiment of Rose Valley Farm and Rochester PJEC. Support ecological agriculture by signing up for six months of farm-fresh organic vegetables. Receive a family-sized variety pack once a week, delivered to a central pickup point in Rochester. Volunteer to help make the farm work by doing a share of harvesting, packing, or weeding. Make the farm connection—don't pay the grocer!

At the meeting, we emphasized that this was to be an experiment. To make it work, everyone would have to be flexible, willing to participate, and ready to readjust. All twenty-four households signed up and made the commitment to share in the labor. As it turned out, a few of the chemically sensitive HEAL members were too sick to do more than make occasional phone calls. The participants filled out vegetable order forms that enabled us to set the contents of the weekly packets. Vegetarian households decided to purchase two or more packets, bringing the total to thirty-one. We agreed that everyone would receive the same selection, except for members on macrobiotic diets, who could substitute turnip greens and collards for tomatoes and eggplant. We set the fee on a sliding scale of $5 to $7 a week for twenty-five weeks, a modest amount of money, but at the request of the members made the packets much smaller than many other CSAs. We designed a "food unit" for a two-person city household. Rather than requiring a lump sum payment at the beginning of the season, we asked for monthly payments in the hope of making CSA affordable for the lower-income members. Social justice has always been high on the agenda of the Politics of Food, a commitment shared by the people at our founding meeting.

We learned a lot from our mistakes that first season. The distribution of the food was the biggest challenge. The members who worked at the farm on each harvest day transported the food the full hour's drive into the city of Rochester, where most of the members lived, and stored it in the coolers at a church. At six o'clock in the evening, two assigned members came to the church to separate the produce into shares. Even with a posted description of the process in excruciating detail—one that would have been the envy of any technical writer—the distribution did not always go smoothly. We concluded that distribution needed a trained coordinator. One person, Jamie Whitbeck, shouldered the task of scheduling all members' work for the entire season. Jamie did an incredibly conscientious job and burned himself out. We learned that big jobs need to be shared by several people. Jan Cox, the bookkeeper-treasurer for the project, eventually collected all the money due, but not without repeated reminders, calls, and cards to the tardy. The next year, we instituted a contract with a commitment to a definite weekly fee and a clear payment schedule. We also stopped allowing members to cancel their shares if they went away for a week or two, which had meant that every week the number of shares was different.

We held a dinner to celebrate the end of the season, the beginning of a lovely annual ritual. The con-

Genesee Valley Organic CSA
Rochester, New York

sensus of the members present was that the experiment, despite a few organizational flaws, was worth repeating. All but three of the families signed up for the second year.

In his 1993 survey of CSAs around the country, Tim Laird found that farmers initiated 79 percent, farmers and consumers together 6 percent, and consumers alone 5 percent of CSAs. . . . Prior to forming these CSAs, 49 percent of the farmers were not farming. Few of them had jobs even remotely connected with agriculture.

AN ARTICLE ON THE KIMberton CSA in *Mother Earth News* in 1989 inspired Jean Mills and Carol Eichelberger to turn their Alabama farm into a CSA. The two women had been partners in a non-farm business for seven years, while living on the farm where Jean grew up. They wanted an excuse to stay at home on the farm and to expand their organic garden, so they shared the article with friends and business contacts. Twenty attended an organizing meeting and eighteen signed up. By the time the 1990 growing season began, the project attracted twenty-nine more. Most of the members were professional people from Tuscaloosa, a nearby university town, joined by a sprinkling of old-time residents. They discussed and settled on share cost, crops, and a distribution system. To get more information about running their CSA, Jean and Carol attended the first CSA conference in Kimberton, Pennsylvania that winter. "It's scary when I think of how little we knew when we started," Carol admits. But they survived that first season and grew to seventy members the next year, and have capped membership at one hundred for the last seven years.

AFTER HEARING GREG WATSON, Massachusetts Commissioner of Agriculture, address the devastating effect the globalization of the food system has had on local farming, Sarah Lincoln-Harrison and her husband Richard Harrison wanted to do something. With five other families in Marblehead, Massachusetts, they decided to find a local organic farm to support. A suggestion from Lynda Simpkins, manager of the Natick Community Farm, led them to Edith Maxwell's one-acre farm, an hour's drive away. For the 1993 season, they contracted with Edith to pay $390 a share and to pick up the bagged produce at her farm. Edith included in the bag a list of the contents and a recipe or two. "It was easy to start," Sarah explained at a session of the Northeast CSA Conference in 1997, "without any special philosophy, we just sort of slid into it." At the end of that first season, they wanted to expand the membership. Since Edith could not grow more, she suggested another farmer, Dick Rosenburgh, who had twelve years of experience growing organic produce for farmers' markets and could handle the potential for growth.

THE FOOD BANK FARM CSA began as a project to produce vegetarian chili for the Western Massachusetts Food Bank. After ten years of supplying government surplus and donated (mainly processed) foods to low-income people, the Food Bank decided to try to produce some of the food itself. Michael Docter raised $10,000 grant money and grew the ingredients for the chili on land at Hampshire College. "It was a great idea," Michael said at a workshop at the winter 1998 PASA Conference, "but no one liked the chili, and it was too inefficient."Michael, having heard about CSA from Robyn Van En, visited a few, decided to adopt that model and scale it up. He wrote a brochure, sent it out to the Food Bank mailing list and local environmental organizations, and plastered copies on bulletin boards. One hundred families signed up for the first season.

Meanwhile, Michael was searching for land on which to farm. The first time around, the land he eventually selected seemed like much too big a piece. In the succeeding months of the search, farmers opened his eyes to the use of tractors for growing on a larger scale. With the backing of philanthropist Ralph Taylor, the Food Bank was able to borrow $230,000 to purchase an old farm on prime Hadley loam soil and to persuade the Commonwealth of Massachusetts to sweeten the deal by acquiring the development rights. With support from CSA members, Food Bank contributors, and grants, they paid off the loans in only three years.

FARMING MORE THAN THREE HUNDRED ACRES on the outskirts of San Diego, California, Bill Brammer has raised mainly tomatoes, squash and cucumbers for the wholesale market. For California, Be Wise Ranch is a middle-sized operation. Bill has been watching nervously as the big boys—Cal Organics, Bornt Family Farm, Pavich—lower wholesale prices by expanding production in Mexico, where they exploit the cheaper land, water, and labor. He worries that wholesalers and distributors of organic produce are ready to source wherever prices are lowest. The CSA that Bill added to his other markets in 1993 as a community service has become a significant component of his strategy to survive the competition by switching as much as possible to direct sales. Solely by word of mouth, membership in his subscription-style CSA has grown from forty to eight hundred. A single newspaper article in 1997 brought fifteen hundred calls to the farm, which had to start a waiting list. Bill is confident that, when he has the infrastructure ready, he will be able to recruit two thousand members.

AT THE OTHER END OF THE SPECTRUM, Karen Kerney, the artist for this book, had been sharing the surplus from her half-acre garden in Jamesville, New York, with a few close women friends for several years. In 1993, they felt a mutual need to formalize their food connection as Karen's CSA. That year, four women gathered weekly, starting in early spring to work beds, harvest spring greens and roots, plant seeds, and transplant. When cultivated crops were ready, they harvested together and divided everything four ways. Gradually, over the next few years, the group grew to five members, who each contributed $250 or more to cover garden expenses. Karen's goals for this project did not include making a living. In her words, "I wanted to share the sense of abundance: there is *plenty* of food . . . take what you need! I wanted to buck the existing system that thrives on scarcity, and to demystify the process of growing food." The members learned how to can, freeze, and dry food for the winter, and all now have gardens of their own, so Karen has reduced her role to growing corn and winter storage crops with her partner John Sustare.

THE DECISION TO FORM A CSA

The image of the ideal CSA is tremendously alluring, the antidote to so many of the ills of the dominant global supermarket. People who have never even gardened hear of CSA and decide they want to become farmers! The dream CSA is a smoothly functioning organic or Biodynamic farm dividing up all its produce among a committed group of supporters who share with the farmers the risks and benefits of farming. With a market assured and income guaranteed, the farmers can concentrate on producing high-quality food and practicing careful stewardship of the land. The members get to eat the freshest, tastiest, most nutritious food they have ever experienced, as though they were master gardeners, but with much less work. They and their children learn fascinating lessons about food production and, by eating seasonally, make a deep connection to a very special piece of land. They respect and honor the farmers' skills and hard work and express their appreciation through friendship, financial support, and helping on the farm. Members and farmers converge into a vital and creative community, which celebrates diversity, both social and biological, and makes food justice and security a living reality. Local, regional, and even international networks of CSAs and other sustainable food enterprises supply members year-round with ecologically produced and fairly traded foods.

That was Robyn Van En's vision and what got her out of bed in the morning. Each one of us, farmer or consumer, can decide to take a few steps toward this ideal. Although none of the existing CSAs has reached it, every single one has achieved at least some small part. Farmer or non-farmer, we need to think through carefully when and how we might want to get involved.

The New Town Farms CSA brochure provides an excellent summary of most of the arguments in favor of CSAs (Randy Treichler has an even longer list in *The CSA Handbook*, pp.13–18).

The Good Fruits

In the community-supported farm structure, every member of the relationship benefits: the shareholders, the farmers, the farms (the Earth), and the greater community.

The Shareholders

- receive fresh, contamination-free vegetables and herbs delivered on the day of harvest;
- pay close to supermarket prices for fresh, certified organic produce;
- know where and how their food is grown, who grows it, and have the opportunity to partake in the miracle of growing food;
- are provided with a structure through which they can support a viable local agriculture, preserve local farmland, and contribute to a healthy local economy;
- have the opportunity to gain knowledge of growing food and stewardship of the Earth;
- become more aware of their relation to the land, farm life, and processes that make our lives possible.

The Farmers

- are given the opportunity to make a viable income by growing food in a responsible and harmonious way, directly supported by the consumer—no middleman;
- have the pleasure of knowing who their product is going to and consequently feel more care, responsibility, and reward in their work;
- are relieved of marketing labor and can focus more on growing food.

The Farms

- are preserved from development;
- are preserved from harmful farming practices;
- are nurtured into a fertile, bountiful land.

The Greater Community

- benefits by the preservation of open spaces, and the maintenance of an important agricultural component that is rapidly being consumed by development and industry—by preserving this diversity the community becomes a more whole and satisfying place to live;
- is strengthened by the bringing together of people who share healthy concerns about our future;
- gets an economic boost when food dollars remain within the community rather than supporting out-of-state corporations.

To recruit members who will stay, it's also important to present an honest picture for people who may choose not to join. In "The CSA Connection," Vicki Dunaway gives a good summary of aspects that will scare some people away:

CSA involves responsibility on the customer's part—to pay up front for the food, to pick up the products on an assigned day, sometimes to return baskets or other containers, and to accept some risk for failed crops. If it's a bad tomato blight year, you may not get all those ruby red tomatoes you have been dreaming of. You might find a corn earworm in the tip of your corn ear if you prefer unsprayed produce (country folks just cut off the tip and never know the difference). There might be a little soil left on your potatoes because they keep better when they are not scrubbed. Often shares are not customized, and you must have an open mind about the type of produce you will receive and be willing to try new things. You may be required to attend a meeting or to work off part of your share by spending a few hours at the farm over the season.

Converting to CSA is easiest for farmers who own a piece of land and some equipment, have a few years of experience growing vegetables for market, and have established a following of customers. A farmer in this situation still needs to think through carefully whether a change in marketing and increased involvement with customers is desirable. Among the many questions to consider: Will growing for a prepaid group of customers be more of a benefit or a heavy obligation? Is my farm located in an area where I can attract CSA members? How will this change my cropping systems? How many shares can I handle? Do I want to combine elements of CSA with other marketing? How much participation from members would I like on my farm? Will I enjoy going to more meetings, or is there someone else who can represent the farm? Will this help to stabilize my income, or just add a lot more work servicing members? Does this fit in with my long-term goals for my farming and my life?

Razzll Wood Farm CSA
Tiffin, Ohio

The obstacles multiply for farmers without land and would-be farmers without experience in growing for market. Yet around the country, many people have found ways to get started once they made up their minds. Sarah Sheikh had a few years of experience as an apprentice and lots of community contacts in the Trumansburg, New York, area where she grew up, but no land. She was able to persuade farmer Tony Potenza to allow her to use some of his land and equipment to start a CSA. Deb Denome, Sally Howard, and their friends had access to some land, a lot of organizing skills and energy, but no farming experience. They advertised on the Internet and found Woody Wodraska, an experienced grower who was willing to take up the challenge of helping them get started on unbroken pastureland. The Acorn Community, in Mineral, Virginia, which already owned some farmland, spent three years getting its gardening, business, and organizing team in shape before launching a CSA. Hundreds of young would-be farmers are finding intern and apprentice opportunities on CSA farms, where they can learn the trade and see if they like it enough to make a life commitment to farming. On more and more college campuses, from Cook College in New Jersey to the University of Montana in Missoula, groups of students are linking up with teachers of sustainable agriculture, food systems, or nutrition, and using school resources to form CSA farms or gardens. (See chapter 21 for a profile of the Cook College Student Organic Farm.)

For non-farmers, the decision may be as simple as finding a local CSA. Several organizations maintain databases on CSA farms by state or region. You can call the Biodynamic Farming and Gardening Association at 1-800-516-7797 to request a list for your state. CSA West (916-756-8518) has an up-to-date list-

Why I Farm

Did last week's newsletter make you wonder why John still farms?

I farm because . . . well, er . . . it's kind of hard to say.

I wanted to write about Neil Diamond impersonation this week, but Kimberely thought it was too much like last week's theme. She suggested I write about why I farm. Then a neighbor showed up. His speaking fit this topic well.

The neighbor said, "I got the arthritis bad, but why wouldn't I after thirty years of beating up this body—broken bones all over. Broke my ribs twelve times working with those cows, broke both ankles, dislocated my shoulder, had to milk with one arm in the air. Whatcha' gonna do? Cows gotta be milked. Couldn't get any help. We offered sometimes up to ten dollars an hour, and we couldn't get kids to show up more than two days. You gotta get the cows milked. It just got so I did it myself—I didn't care what was busted."

Our neighbor didn't say exactly why he farmed. That's not a very farmerly way to talk out here. But I noticed that there was something he liked very much about farming, or he wouldn't have been doing it.

Another farm family nearby is legendary for getting their crops in first. They move fast, all ages, in a spritz of tobacco juice and beer. Several of them are missing toes and fingers from machinery accidents. One limps from a hip smashed by a bull. The last finger that got lost didn't even stop the haying.

It's hard to explain just what causes a person to stay in such a life. For me, as I miraculously type with all ten digits, I think about when I suddenly went from a fleet of cars and trucks and an arsenal of machinery down to nothing in the early eighties. My boots were worn out, and I didn't have the money for another pair. My mother bought me a pair. I will forever remember the exquisite sensation of walking what was left of this farm, secure in my shiny rubber boots, feeling somehow that those boots had restored me to the land. The land has a feel underfoot that can melt a person to it. And then there's the smell—our machine shed has a smell of eternity, a musty ancient fragrance from before my birth and into the hereafter. There's the rhythm—the barn door opens and closes; the swallows return; the brome grass swishes.

On NPR, Susan Stamberg interviewed a Mayan girl in the Yucatan Peninsula (through a translator). She wanted to know why the girl weaved all day long. The girl didn't answer to Susan's satisfaction, so Susan kept trying to put words in her mouth.

"Is it because you can sell your weavings for money?"

"No."

"Do you weave because your ancestors weaved and it's a way to stay connected to your people?"

"Huh?"

"Do you weave because you love the rhythm and the patterns of weaving?"

"No."

"Why do you weave, then?"

"I just weave."

I don't stay on this farm because brome grass swishes. That's a fringe benefit. The closest I can describe my bond to it is a shudder I get, an irrepressible vibration when it's time to work the fields. I can be eating, sleeping or talking on the phone, and when the time is right to plow or plant, my body registers some mysterious sensation, an irresistible beckoning. My legs take me to the work, put me on the tractor; I am all surrender. And the joy of pushing dirt around, the ecstasy of spraying potentized silica, the thrill of organizing little dots of green into straight lines on bare soil—these invoke in me a subtle delirium.

For two years I toured rural Illinois with a play I wrote about a farm family losing its land. Audiences wept and laughed. Once an old man caught up to me backstage. He said, "Let me tell you how to farm. There's only one way. You farm 'til you're down to your last nickel. And then you keep farming until the nickel's gone."

Boundaries? Where are the boundaries here? I don't know. Like a drug, the land can lure a person into destitution. It can overshadow one's love for others. The land can embolden, exhaust, ennoble. It can nurture, destroy, sustain.

I don't think I have boundaries with the farm.

I just farm.

—John Peterson, Angelic Organics, Caledonia, Illinois. Used by permission.

ing of California projects; the Michael Fields Agricultural Institute (414-642-3303) rides herd on the Upper Midwest; the Northeast Sustainable Agriculture Working Group (NESAWG) office (413-323-4531) has taken over the Northeast from CSANA; and the Southern SAWG (501-587-0888) keeps an eye on the South. Where no CSA exists, people like the Harrisons (see p. 27) or Peggy and Martin Danner, in Boise, Idaho, have reached out to other consumers and local farmers to start new ones. The Danners founded the Boise Food Connection, a nonprofit dedicated to facilitating CSA in the Treasure Valley.

Converting to CSA is easiest for farmers who own a piece of land and some equipment, have a few years of experience growing vegetables for market, and have established a following of customers.

In his 1993 survey of CSAs around the country, Tim Laird found that farmers initiated 79 percent, farmers and consumers together 6 percent, and consumers alone 5 percent of CSAs. Two farms reported that organizations had started them and one farm began because of the work of a student intern. Prior to forming these CSAs, 49 percent of the farmers were not farming. Few of them had jobs even remotely connected with agriculture. Previous jobs they listed included postal worker, agricultural consultant, plumber, laboratory technician, health club owner, teacher, truck driver, social worker, student, journalist, engineer, veterinary clinic manager, produce buyer, nurse, and underwater photographer. Five farmers went from part-time to full-time to run CSAs. (Laird, p. 45; Tim has graciously allowed me to use his work as a source of information for this book.)

Farmers considering the move to CSA should expect changes in their work patterns. In Laird's survey 67 percent of all the farmers felt their work load had increased, though only 55 percent of those with previous experience said this. For 30 percent, the load remained the same, while only 3 percent noticed a reduction. Of the thirty-four new farmers, 85 percent said that their work hours had increased. The need to grow a greater variety of vegetables in order to make up a good share, and the organizational demands of running a CSA created the extra work. Tim Laird observed from his own experience in market gardening as compared to CSA that the members' expectations for quality and variety made production for the CSA more stressful. On the positive side, 79 percent of Tim's respondents said that their job satisfaction had increased, 17 percent said it remained the same, and only 4 percent felt a decrease. The two new farmers whose job satisfaction diminished drew the logical conclusion and left farming, as did the one experienced farmer. CSA, clearly, is not for everyone, but for farmers like Carol Eichelberger in Alabama and David Inglis in Massachusetts, it is the only way they can imagine going on with farming. Dorothy Suput wrote in the summary of her study of CSA farms in Massachusetts:

> All farmers participating in a CSA organization commented on the reduction of market-related stress [leaving] extra time for training apprentices, educating members, socializing, spending time with family, organizing agricultural education programs, and giving extra care to farming. (Laird, p. 50)

Larry Halsey told me that switching from selling at Greenmarkets, the farmers' market in New York City, to a CSA with two branches in Brooklyn and Queens, reduced his workday from eighteen hours to six and improved his family life. In the 1996 season, the first year GVOCSA was the primary market for Rose Valley, I no longer had to spend several hours a week on the phone selling vegetables to produce managers. (There are almost no contracts in fresh market vegetables, so there is always the nagging worry that one week no one will want what you have to sell.) Our farm team did not have to pick hundreds of bunches of greens and herbs on a tight schedule. No one from the farm had to load the pickup truck, rain or shine, with heavy boxes and bags of vegetables and drive the hour or hour and a half to deliver to stores in Rochester and Ithaca. There were no emergency phone calls from apprentices asking how to change a flat tire or what to do about a truck breakdown out in the middle of nowhere. The police never called me to ask if I knew the man named Greg they held in custody.

Growing food for the GVOCSA had a more relaxed rhythm to it. On Tuesday and Saturday mornings, I walked the farm to see what was ready to put in the packets for the next day. I assembled the bulk order, filled out instructions for the distribution crews, and made plans for the work crews. Many Wednesdays and Sundays, we started picking at seven to be sure to get the tender greens in the cooler before the sun cooked too hot. As CSA members arrived, we slowed down to a more social pace. Those mornings were an intense mixture of leisurely personal encounters and goal-oriented work. Our sense of responsibility was matched by the remarkable conscientiousness of the members.

Describing the process of growing food as "life-transforming," Karen Kerney says, "the best part of the CSA was the opportunity to focus on growing food that I knew would be picked and consumed by people who cared as much as I do about how our food is grown. Relieved from the incredible amount of energy it takes to harvest, package, and market food, I was able to center my energy on the art, science, and craft of growing food."

In his poem "rural population base/distance from/stretched out possibilities," Paul Bernacky of Wayback Farm in Belmont, Waldo County, Maine, writes about how his CSA with twenty summer and winter shares, in its eighth year in 1998, changed his work:

> Organize with a purpose living
> in balance and harmony transferred
> stress into dynamic tension rekindless
> my sense of wonder.

STEPS TO FORMING A CSA

Whether consumer- or farmer-initiated, the steps needed to begin are similar. (See the chart on p. 34 for a summary of the process.) First, find out if enough people are interested. You can do that very informally, by polling your friends or customers, or more formally by calling a meeting. The Danners posted this engaging meeting announcement:

What's special about these vegetables? They're fresh, local, and grown for you! If you are interested in any of the following:

- A steady supply of fresh food grown locally
- Knowing the people who produce your food
- Providing direct input on what food is produced and how
- Fixed prices established before the growing season
- Trying different types of produce grown locally, along with information on how to prepare them
- Supporting local, small-scale farmers
- Participating in a grassroots effort to directly connect local growers and consumers

Then contact us for more information or attend our first planning meeting.

That first organizational meeting should set the tone for the entire project. If you want to run things and do most of the work yourself, you can present

Steps to Forming a CSA
(Start small and grow organically!)

1. **Initiators (either farmers or group of non-farmers) issue a call to form a CSA. You can seek members**
   a. among friends or neighbors
   b. among existing groups: daycares, environmental or consumer organizations, churches, civic groups, schools or other institutions, workplaces

2. **Hold exploratory meeting of prospective sharers and farmer(s). Possible agenda:**
   a. what is a CSA?
   b. why eat locally grown food?
   c. why small farms need support
   d. assess level of commitment of participants
   e. if interest is high enough, create founding core group

3. **At this meeting or a subsequent meeting, come to agreement on the group's values**
   a. does the group want organic food?
   b. does the group want locally grown food?
   c. does the group want racial, ethnic, and economic diversity among members?
   d. is it important to involve children?
   e. will all members contribute work, or will some buy out by paying a higher fee?
   f. do members want to share production risks with the farm(s)?
   g. what commodities does the group want?
   h. does the group want to share mailing list with other groups?

4. **Organize the core group to**
   a. decide on farmer(s)
   b. decide growing site
   c. decide how and where food will be distributed
   d. divide up member responsibilities
   e. approve the budget proposed by the farmer(s)
   f. set fee policy and payment schedule
   g. clarify expectations as to variety and quantity of food
   h. set guidelines on participation of children (if desired)
   i. decide who owns any equipment purchased

5. **The core group recruits members for first season**
   a. post fliers
   b. organize recruitment meeting
   c. talk up idea with friends
   d. place notices in organizations, churches, and do flier mailing to likely groups
   e. send out press release
   f. find friendly reporter to write story

6. **Members make commitment**
   a. to pay in advance of receipt of food (whether by season, month, or other schedule), and regardless of quantity and quality of food due to weather conditions
   b. to participate in farm, distribution, and other CSA work

7. **Establish the legal status of the CSA. Many groups defer decisions on legal structure for a season or two. Advice from a lawyer may be helpful. Existing options include:**
   a. consumer cooperative
   b. sole proprietorship or partnership of farmer(s)
   c. corporation or limited liability corporation
   d. nonprofit corporation (or branch of existing one)
   e. farmer-owned co-op

8. **Determine capitalization of the farm(s). Many CSAs start with minimum of rented or borrowed land and equipment. For the longer term, decisions must be made on purchase and maintenance. Options include:**
   a. farmer(s) capitalize
   b. members capitalize through fees
   c. the group seeks grants
   d. the group seeks loans. Possible sources include Farm Credit, National Cooperative Bank, members, commercial banks, revolving loan funds, pre-sale of farm produce coupons

   Options for land tenure include:
   e. private holding
   f. land trust
   g. lease agreement with private owner or institution

the information about your project to the participants and invite them to sign up. If you hope to make the project more participatory, however, begin by giving everyone at the meeting the chance to talk about why they came, what they expect, and what they plan to contribute. Begin the agenda with some discussion of what CSA is, its history, and what is going on locally. Participants need to understand why it is important to eat locally grown food and how the dominant food system endangers local farms. After you've presented some historical and philosophical perspective, you can get down to brass tacks. Is there enough energy and commitment to start a CSA? Who wants to get involved and what specific jobs are they willing to do? If you can come out of this meeting with the beginnings of a core group, you have done a good piece of work.

Valley Creek Community Farm
Northfield, Minnesota

At this meeting or the next, you need to agree on the values for your CSA and the policies that flow from those values. Do members want organically or Biodynamically produced food, or are they satisfied to support a farm using any practices as long as it is local? Does the group want to foster ethnic, racial, economic, or age diversity among members, or contribute food to the hungry? Should children be involved? Besides food, will there be an educational component? Will the project ask every member to work or will some members be allowed to exchange labor for a lower price? Do members want to share production risks with one farm, or should provisions be made to purchase from several farms? What commodities will the CSA distribute? Once these broader questions have been answered, you can tackle the details of how the CSA will function. What farm or farms will produce the food? How much food will go in a share? What kinds of shares will you offer? How much will a share cost? Where will members pick up their food? With this information in hand, you can draw up a flyer or brochure and start recruiting members. (See sample brochure on p. 38.)

A farm that has customers already has a good base with which to start. Many CSAs have grown mostly by word of mouth, and the new members who learned about the CSA from old members are more likely to have realistic expectations. Dru Rivers of Full Belly Farm in California says that 80 percent of their new members come from word of mouth. Every year, GVOCSA sends out packets of three to four brochures to current members and asks them to recruit their friends. Gradually, some entire workplaces have signed up. One member who is a nurse served Rose Valley salads at hospital staff meetings and attracted a dozen or so other medical personnel. A young resident lured in the other members of his class. Caretaker Farm was able to convert totally from other kinds of marketing to CSA in one season because of their established connections and excellent reputation; 125 people attended their founding meeting in September 1990. Few farms, however, could attract that many potential members all at once.

The best way to find members is to approach active groups of people who are already organized around some related concern: schools, daycares, church groups, environmental, political, or farming organizations, or workplaces. The GVOCSA started with members of POF and HEAL, many of whom had eaten Rose Valley vegetables sold at the Genesee Food Coop, and expanded by recruiting members of the Sierra Club and Corpus Christi Church. The Kimberton and Brookfield Farm CSAs grew out of close relationships with Waldorf schools. Members of the Albany Catholic diocese, who had been inspired to take action in support of local farms at a New York Sustainable Agriculture Working Group meeting, approached Roxbury Farm with an offer to form a group. Joe D'Auria founded a CSA with members of the Long Island Chapter of the Northeast Organic Farming Association.

People with chemical sensitivities, cancer victims, and families with autistic or highly allergic children are attracted to CSAs by the hope of alleviating their illnesses. Helping members improve their health can be very gratifying. One member family of the GVOCSA told me that their autistic son was calmer and easier to live with when he ate our vegetables. Unfortunately, the fresh produce therapy does not always work or is started too late: our core group suffered together through the final months and eventual death from cancer of our treasurer, Joe Montesano. In a way, I guess, that is also one of the risks of farming for a community.

The workplace is another source for membership. Cass Peterson and Ward Sinclair sold shares in the Flickerville Mountain Farm and Groundhog Ranch to their former colleagues at the *Washington Post*, delivering to the newspaper office building once a week for six years. Rodale employees make up most of the sharers for George DeVault's CSA. In 1995, our neighbor Glenda Neff signed up seventeen sharers at her workplace an hour away in Syracuse. This mini-CSA lasted as long as she kept the job and did the weekly deliveries. A member of the board of Nesenkeag Farm offered his corporate contacts as a way to recruit some shares. A total of thirty people in two offices signed up. Farmer Eero Ruuttila delivers the weekly shares packed four or five to a styrofoam cooler box so that the produce stays fresh and does not take up too much space in the office buildings. He limits the size of the shares to what can be carried home easily in one plastic market bag. Communications with members was simplified by office e-mail.

Dan Guenthner, who farms with his family at Common Harvest Farm near Minneapolis, Minnesota, has proposed Congregation Supported Agriculture. In "To Till It and Keep It: New Models for Congregational Involvement with the Land" (1994), Guenthner states his view that "the Bible is nothing short of an agricultural *handbook*!" He outlines sixteen ways in which parishes can connect with local farms, better educate their members about local food systems, and generally empower them to participate in land stewardship. His booklet, funded by the Evangelical Lutheran Church, gives several examples of church-sponsored CSAs. Support from local churches enables Common Harvest Farm CSA to distribute 15 percent of its shares to low-income families. The response will differ from parish to parish, but local churches with active social justice committees are fertile ground for would-be CSAs. (See Guenthner's list of suggestions for linking CSAs and churches on p. 37, and "Faith Communities on the Land" by Patricia Mannix on p. 206.)

You should not have to pay for advertising. Brainstorm a list of likely places to display your brochure: health food stores, food co-ops, and vegetarian restaurants; health clubs, chiropractic clinics, or other alternative medical offices; recycling, Yoga, tai chi, or spirituality centers; and church and school bulletin boards. Local radio and TV talk shows are often hungry for interesting people to interview. Find a literate reporter for a local newspaper and ask him or her to write a story about your CSA. Attracting the attention of major newspapers may be difficult, but small-town papers are usually anxious for a good story with photo opportunities. Earthcraft Farm signed up twenty new members after an article in a local paper. And even the big rags have given CSA some attention. For years after Ward Sinclair retired from its staff to become a farmer, the *Washington Post* published his "Truckpatch" column. During the summer of 1997, the *New York Times* food editors split a share in a CSA and wrote a whole series of stories. Even though the tone of most of these stories was rather snippy and showed a remarkably low level of understanding of what CSA and eating in season is all about, they attracted new sign-ups for the CSAs mentioned.

When potential members do call in response to an ad or a flyer, have a seasoned member available to answer their questions. The office at Angelic Organics keeps a list of members who are willing to talk to new recruits. See the example from Wishing Stone Farm of an attractive flyer on page 38.

At the 1997 Northeast CSA Conference, Nancy Schauffler, an active member of Stoneledge Farm CSA in New York, listed seven recruitment techniques

What Can We Do to Respond to the Needs of the Land?

*by Dan Guenthner**

- Introduce the concept of congregationally supported agriculture in your church.
- Plan to take at least six to nine months to form a core group of interested members to help organize and set up the details of a marketing relationship with a local grower.
- Find existing vegetable, dairy, fruit, and meat farmers in your area. Contact them directly to see if they would be interested in participating in a direct partnership with your church.
- If local farmers are not available, consider providing an entry-level farmer/gardener with the opportunity to get established through the benefit of an immediate market guarantee.
- Church members may have land available for a farming/gardening enterprise. Investigate and evaluate the existing resources within the church community.
- Churches can provide many initial benefits in starting a CSA—a committed community of people, access to distribution information through newsletters and bulletin boards, and weekly contact with each other. The church can help establish a *fair* price for food.
- Churches can also purchase shares in a congregationally supported farm for other ministries such as the local food shelf, shut-ins, and people with AIDS and other illnesses that prevent access to fresh food.
- Contact your synod or conference office to investigate if other church-owned land might be available for this type of innovative and direct form of soil stewardship.
- Approach your local or regional church camp to see if a church-supported farm can be established on tillable land at the camp. Vegetable cultivation and harvesting can become a meaningful part of the camp activities and summer program.
- Be creative! One of the most exciting things about this new form of food distribution is that each farm is unique and is determined by location, farm size, community, and climate. No matter what form the CSA takes, the expressed outcome is the same: Non-farm citizens now have access to a food supply that is socially responsible and built directly upon their helping to care for the soil.

* Dan Guenthner farms at Common Harvest Farm in Minneapolis, Minnesota. Contact the Land Stewardship Project (listed in CSA Resources) for more information on congregationally supported agriculture.

WISHING STONE FARM

Organic and Specialty Grown Produce

Now Offering 1994 Community Supported Agriculture Memeberships

25 Shaw Road
Little Compton,
RI 02837
(401) 635-4274

- Organically grown vegetables
- 9 Month season; April thru December
- Farm root cellars keep Carrots, Beets, Potatoes fresh through winter season
- Summer season; Fresh Flower Bouquets, Fresh Baked Bread, Herbs, Local Dairy Milk and Organic Beef
- Heirloom vegetable varieties and European cultivars to spark your imagination

Flexible pickup schedules:
Summer- 5 Days a Week
Spring and Winter pickup by appointment

What Is Community Supported Agriculture?

Community Supported Agriculture (CSA) is a consumer based movement which allows families to directly access farm fresh certified organic produce, while simultaneously supporting sound environmental practices and preserving open space in R.I.

REQUEST INFO:

WISHING STONE FARM
A Rhode Island Certified Organic Farm

Community Supported Agriculture (CSA)

Organic Produce, Fresh Cut Flowers and Herbs, Fresh Baked Bread, Local Dairy Milk and Organic Beef

Flexible pickup schedules, daily or monthly..
Prepaid in the spring and enjoyed all Summer and Fall

they tried the first year, and quantified how many members each brought in:

1. an existing parent-teacher-student environmental group—6
2. a mailing to Just Food members, who already grasped the concept—5
3. posting flyers with a tear-off phone number—20
4. talking to everyone you know—20
5. notices in organizational newsletters—5
6. an ad on public radio and an article in a newspaper—lots of members for another CSA!
7. recruitment meeting—the meeting gave the sign-up campaign a deadline to work for and solidified the group already recruited. She gave additional pointers: "It gives people the chance to meet the farmer, the core group, and each other. Have lots of literature available, consider showing a video, and have contracts there so people can sign up on the spot. Pass around an attendance sheet, and ask people to write areas in which they would like to volunteer, specifying newsletter, accounting, publicity, recipe collection, and the like."

Two videos and two slide shows introducing CSA are available for meetings of this kind. In 1986, Robyn Van En, Jan Vandertuin, and Downtown Productions of Great Barrington, Massachusetts made "It's Not Just About Vegetables," which is now a classic. More recently, the Center for Sustainable Living CSA project at Wilson College produced a video, "CSA: Making a Difference," and two versions of a slide show, "CSA: Be Part of the Solution." One version supplies text slides only and allows you to fill in your own farm slides, while the other includes slides of CSA farms as well. These materials are well made and inexpensive. (See the CSA Resources section for information on Wilson College and other CSA support materials.)

Among the more creative promotional ideas used by CSAs: John Clark, of Clark Farm CSA in Connecticut, and David DeWitt, of First Light Organic Farm in Massachusetts, give tours of their farms to schoolchildren whose families then sometimes join. As a group, the CSA farms in Vermont offered sample shares as a benefit for supporting Public Radio. While this did not generate many new shares, it got CSA favorable radio coverage. CSAs in New York City attracted a lot of attention by holding a tomato tasting with 160 varieties of tomatoes at the public market. West Haven Farm at Eco-Village in Ithaca, New York, asks members who are going away for a week or so to give their shares to friends, who then often sign up. The Partner Shares Program of MACSAC sponsors "farmathons" in which volunteers solicit financial sponsors for their work on farms and the proceeds go to purchase shares for low-income members.

Research projects, such as those conducted by Gerry Cohn, Deborah Kane and Luanne Lohr, Rochelle Kelvin, Jane Kolodinsky and Leslie Pelch, and Dorothy Suput, that have investigated why people join CSAs all show that members want fresh food from local farms. Some surveys show organic food as the top reason, others show fresh or local first with organic second. Support for local farming, environmental and health concerns, quality of produce, and food safety come close to the top of the lists. End-of-the-season surveys show that members stay on because of the quality of the produce, with farm work and newsletters as close contenders. Meeting members' expectations is critical to retaining them as sharers in a CSA. (See chapter 12 for more on member retention.)

REGIONAL CSA SUPPORT GROUPS

"Cooperation among cooperatives," one of the six basic principles of cooperatives around the world, is valuable for CSAs as well. Instead of competing with one another, groupings of CSAs and CSA supporters have been coming together to work for mutual benefit and CSA promotion. Like the CSAs themselves, no two of these efforts take exactly the same form. The Madison Area CSA Coalition (MACSAC) in Wisconsin encompasses the farms that service the city of Madison. The Michael Fields Agricultural Institute publishes the "Upper Midwest Regional Directory" of CSAs in twelve states. A dozen or so farms in western Massachusetts and eastern New York, many of them CSAs, share resources in training apprentices through the Collaborative Regional Alliance for Farmer Training (CRAFT). Just Food in New York City recruits city-dwellers and matches them up with farms in the metropolitan area. With the support of a SARE grant, CSA farmers in Iowa have formed the Iowa Network for Community Agriculture (INCA) and hired CSA farmer Jan Libbey to act as education coordinator, and the Iowa State Extension distributes a statewide list and an excellent booklet entitled *Iowa CSA: Resource Guide for Producers and Organizers*. As part of a project to create local jobs and businesses in Appalachia, the Clinch-Powell Sustainable Development Initiative is helping start CSAs in Virginia and Tennessee. CSA West publishes a listing of CSAs in California and organizes informational workshops on CSA around the state. Since Robyn Van En's death and the loss of CSANA, the NOFA Interstate Council has been pooling resources with the Northeast Sustainable Agriculture Working Group and the Wilson College Center for Sustainable Living to sponsor a similar service in the Northeast, while the Biodynamic Association maintains a database of CSAs by state for the entire country. The CSA Resources section lists contact information for these and other organizations.

The success of MACSAC in building CSA membership to over sixteen hundred in a fairly small, if exceptionally progressive, region proves that synergy beats competition. As Marcy Ostrom tells the story, in the fall of 1992, a group of food activists and farmers from one farm decided it was time to get more Madison consumers eating locally grown food (*Farms of Tomorrow Revisited*, pp. 88–89). Instead of forming one CSA, they did an energetic media campaign to publicize the concept of CSA and organized an open house where eight farmers signed up members. Their first CSA fair drew over three hundred people: Madison CSA membership went from zero to 260 in one season. According to John Greenler, the number of CSAs rose to ten the second year, eighteen in 1997 and then fell to fifteen in 1998, when he stopped farming full-time due to back troubles, two CSAs consolidated on new ground, and a third recruited enough local shares to drop its Madison section. In the years since 1992, the open house has become an annual celebration, drawing up to five hundred people to the city botanical gardens with speakers, farm slide shows, activities for children, and membership information on each farm. With the support of the Wisconsin Rural Development Center, MACSAC has evolved into a CSA farming association combining outreach to the public with farmer-to-farmer technical exchanges and mutual support. Similar to MOFGA or the NOFAs, MACSAC holds meetings on practical topics of importance to farmers—such as equipment, vegetable production, and CSA organization—and participates in a regional conference. MACSAC has not tried to define what CSA farms should be, intentionally allowing diversity to take its course as projects mature, and ready to help in whatever way is needed.

Calculating that to match the Madison density of CSA membership in New York City would mean 250,000 member families, Just Food, the NYC Sustainable Food System Alliance, has set a goal of facilitating the creation of thirty to fifty CSAs in ten years. New York City's fortress-like geography, transportation nightmares, and distance from market gardens make this a formidable challenge. Sarah Milstein, a dedicated core member of the Roxbury CSA, the only one previously delivering to New York City, had the job of Just Food CSA coordinator. Robyn Van En inspired and helped in the early stages of this effort, which began with informational meetings for city people in 1994. Borrowing the open-house idea from MACSAC, in the winter of 1996 Just Food's "CSA in NYC" program held a CSA Sign-Up Fair attended by eight farmers and seventy

The Cooperative Principles from Ideology to Daily Practice (The Rochdale Principles)

These seven principles are embraced and utilized by co-ops in all countries of the world. They represent the fundamental features of a cooperative, regardless of the type of business it is involved in. These co-op principles are an important link between the theory of working together and the application of that idea to economic institutions.

These principles were originally formulated by early co-op groups in the mid-1900s and most recently refined and adopted by the International Cooperative Alliance in 1995.

1. **Open and Voluntary Membership**
   Co-ops do not restrict membership for any social, political, or religious reasons. They are open to all persons who can make use of their services and are willing to accept the responsibilities. Members choose to join and participate in the co-op.

2. **Democratic Control: One Member/One Vote**
   Co-op members are each equal co-owners in the cooperative business. Each member has voting and decision-making power in the operation of the business on a one-member/one-vote basis. In this way, no one person gains control based on the amount of money invested or position held.

3. **Return of Surplus to Members**
   Surplus or profit arising out of the operations of the co-op belongs to the members (owners). This is returned on the basis of patronage—or how much business a member did with the co-op (or based on how much the member contributed towards surplus).

4. **Limited Rate of Return on Investment**
   Share capital (the money invested by the member/owners in the co-op as equity) can earn dividends, but it would be at a limited rate. This prevents outsiders from investing money in the co-op to make a profit purely as an investment. The rate of return is typically limited by law to 6 to 8 percent.

5. **Continuous Education to Members and Public**
   Co-ops encourage member participation and involvement in the co-op, educating members in the principles and techniques of cooperation. Co-ops also work to educate the general public and members about their service areas: in the case of food co-ops, about nutrition, food systems, and the like.

6. **Cooperation among Co-ops**
   To complete the cycle of cooperation, co-ops work with other local, regional, national, and international co-op organizations. Co-ops working together help each other and strengthen their individual and aggregate economic positions.

7. **Concern for Community**
   While focusing on member needs and wishes, cooperatives work for the sustainable development of their communities.

potential members. The farmers gave brief presentations about their farms, consumers and farmers matched up by geographical proximity, and core groups began to form. As a result, five new CSAs started in 1996, ranging in size from fifty to over one hundred shares. These CSAs have continued to grow in the years since with some farms adding a second pickup site as well as additional shares. A core group oversees site management and member participation for each site, and most are finding ways to include other food products in addition to the farms' vegetables. CSA in NYC facilitates cross-fertilization through annual meetings of all of the cores and farmers.

Based on their experience with these first five farms, CSA in NYC is planning to establish three new CSAs a year, cultivate more city community groups and eligible farmers in the region, provide technical support for the existing CSAs, and improve linkages between the CSAs and anti-hunger efforts in the city. Sarah Milstein's work with Little Seeds Gardens, a new CSA for 1998, gives a snapshot of how CSA in NYC operates. Two years ago, Sarah met Claudia Kenney and Willie Denner, a young couple who were attracted to CSA farming. Sarah kept in touch with them while Claudia and Willie established Little Seeds Gardens on rented land, learned more about CSA through a year each of membership in the Hawthorne Valley and Roxbury CSAs, and got more experience growing for farmers' markets. Willie says, "I wouldn't want to enter into CSA without this previous growing experience. I am starting with trepidations even after four years of growing vegetables for market."

With her antennae always up for promising connections, Sarah found Jessica Stretton, a Just Food member who wanted to start a CSA in her economically and ethnically diverse neighborhood, an area well located for transportation. Sarah helped Jessica devise a plan: they sent out invitations to thirty-five Just Food and Roxbury CSA members in a few zip codes. (Roxbury has a waiting list and can spare a few members.) Six people came forward to serve on a core group, and twenty expressed an interest in shares. Sarah then contacted Claudia and Willie who said they were ready to do fifty shares in 1998 and perhaps one hundred in 1999. They met with the core group and both sides made a two-year commitment. Sarah supplied the group with a packet of guidelines for the farmers and the core, including forms to help develop budgets. The core members found a neighborhood park to use as a distribution site, and the farmers will host a visit to their farm for the core. Sarah will keep in touch, intervening only with helpful tips, such as setting up a voice-mail box to answer queries, and counseling when problems arise.

A farm that has customers already has a good base with which to start. Many CSAs have grown mostly by word of mouth, and the new members who learned about the CSA from old members are more likely to have realistic expectations.

Meanwhile, Claudia and Willie are hard at work developing Little Seeds in Stuyvesant, New York. For a family with two small children, apprenticeships are awkward, so they have been learning farming by doing it, and taking every opportunity to attend workshops and get advice from the farms in the CRAFT network. Instead of purchasing land, they have put their resources into equipment and slow, but steady growth, from one to one-and-a-half to eight acres under cultivation. They lease half a barn and have purchased a used Farmall Super A tractor, two used hoop houses, a used reefer truck, and a new spader. Despite their small size and inexperience, they are taking on apprentices to whom they can offer the resources of the CRAFT program, which provides tours and training sessions on all twelve member farms. (See chapter 7 for more on CRAFT.) The Little Seeds farmers are pulling themselves up by their bootstraps, but they are doing it thoughtfully, aiming for satisfaction and community in the process.

In harmony with its mission "to increase the availability of healthful, locally grown food to the people of New York City, particularly those with little or no income," Just Food continues to explore new horizons for CSA in the big city. Hard as it is to line up middle-class members for CSAs, Just Food is determined to discover how to make it work for low-income city-dwellers as well. For the CSAs they have helped so far, Just Food has encouraged the offer of working shares and donations of leftover shares to food pantries or soup kitchens. They hope to facilitate planning for larger donations in the future. Sarah and Just Food Director Kathy Lawrence have had some challenging meetings trying to explain CSA to hard-pressed, low-income, grassroots groups that are struggling for survival amidst the harsh economic and social realities of the Big Apple. For white, middle-class organizers, building trust with other ethnic or income groups is a major hurdle to overcome. But Sarah and Kathy have made some breakthroughs and are testing some new approaches, which include:

- holding community fundraisers to buy or subsidize shares for people who can't afford them;
- establishing a revolving loan fund through which the grower could be paid up front, while the members who cannot afford a lump-sum payment pay in installments over the course of the season. If these members move or become unable to pay for their shares, the farmer will not be left holding the bag. On the other hand, if the farm has a crop failure, the fund could cover the risk to members who have spent their limited food budget on the CSA by paying for food from other farms;
- compensating with shares in a CSA the work of a group of parents in East Harlem who are enrolled in a nutrition education program;
- through the City Farms project, creating CSAs for which the food is produced in community gardens in the city.

(For more information about how other groups around the country are pioneering in this area, see chapter 20.)

THE BEGINNING OF EACH NEW CSA is a hopeful moment. The act of selling or buying shares in a farm is full of promise and rich significance. Transforming that act into the living reality of community support, however, takes time and the willingness on all sides to change: to change how we eat, how we think about food, how we pay for it, how we manage a farm, and how farms connect with one another. The more we can share our discoveries, the faster community supported agriculture will be able to grow towards agriculture-supported, sustainable communities. By teaching as we learn, CSA farmers and members may not become millionaires, but we will create solid local institutions and social capital, which gives us strength for the future we cannot even imagine today. Like the Mondragon cooperatives, we are building the road as we travel along it together.

HOW TO CHOOSE A FARMER 4

And he gave it for his opinion . . . that whoever could make two ears of corn, or two blades of grass, to grow upon a spot where only one grew before, would deserve better of mankind, and do more essential service to his country, than the whole race of politicians put together.

—JONATHAN SWIFT, *GULLIVER'S TRAVELS.*

NOT SINCE THE KINGS OF FRANCE selected gardeners based on their ability to supply asparagus year-round have consumers had to give this much thought to the choice of their farmer. Yet choosing the person who will grow your food is surely as signficant as selecting the family doctor, the pastor for your church, or the mechanic for your car.

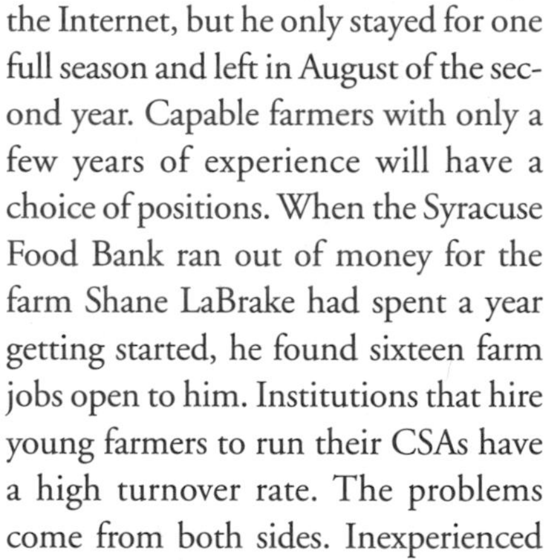

As for any job applicant, be sure to check references. Doing a reference check before interviewing may save time. Prepare a list of the questions you want to ask before calling references and, if possible, find someone not on the candidate's list who can tell you about him or her as well.

To find appropriate candidates, place notices with state organic farming organizations, the Biodynamic Association, and regional CSA centers. AATRA has addresses of organizations and networks that may be able to help (see CSA Resources section).

Unfortunately, you will probably find that few seasoned farmers are available. Most of the older farmers are settled on their own farms. Seeking Common Ground CSA made a lucky catch in Woody Wodraska, a Biodynamic farmer with many years experience under his belt, who they found through the Internet, but he only stayed for one full season and left in August of the second year. Capable farmers with only a few years of experience will have a choice of positions. When the Syracuse Food Bank ran out of money for the farm Shane LaBrake had spent a year getting started, he found sixteen farm jobs open to him. Institutions that hire young farmers to run their CSAs have a high turnover rate. The problems come from both sides. Inexperienced farmers may not be able to cope with the many levels of organization required to run a CSA, while some institutions place additional burdens on staff that are not appropriate for anyone trying to farm. (Kristen Markley analyzes this problem thoughtfully in her Master's thesis, "Sustainable Agriculture and Hunger.")

A few CSAs have gotten started without a skilled farmer. The Bawa Muhaiyaddeen Fellowship, a Sufi community in Philadelphia, managed to keep the Farm Food Guild CSA going for three years with four and then two members taking the lead, and others helping out as they could. With fifty families signed up, Laura DeLind, a professor of anthropology, found herself in the role of a thirty-hour-per-week grower at the Growing in Place Community Farm in Michigan. She had expected that other members would participate more actively and that they would "learn together how our food is grown." The CSA survived the season, but Laura experienced "no joy," intense anger at those who seemed to let her down, and a lot of stress. A career change is not in Laura's future, though she was able to convince the group that is taking the lead for 1998 to hire an experienced organic farmer as a part-time consultant.

Any number of other cooperative CSA gardens are underway, such as the Genesee Community Farm in Waukesha, Wisconsin. "Learning as we grow," is their motto, and in their invitational brochure they emphasize community-building and member participation. With only fifteen shares, perhaps they can avoid the strains suffered by Growing in Place and develop a more equitable division of labor and responsibility. After receiving weekly shares for three years, the five working members of Karen Kerney's CSA in Jamesville, New York, found that they were supplying day-to-day needs from their own gardens. Karen obligingly shifted to field crops of winter squash, corn, and potatoes, and the members reduced their weekly work shifts to occasional work parties at peak planting and harvest times. If they return to weekly harvests in the future, Karen says she will insist on a firm work schedule; trying to accommodate everyone's individual needs became overwhelming. It will be fascinating to see how these noncommercial gardens evolve, and whether they will find a path to long-lasting cooperation.

Conceivably, a group of people could decide to form a CSA in order to give a new farmer the opportunity to get started. If everyone involved understood

Hiring a Farmer

A CSA that has land, but no one to farm it, might consider the following in hiring a farmer:

- **Previous farming experience:**
  appropriate crops
  similar scale
  comparable climate and growing conditions
  apprenticeship or internship on CSA farm
  farming or gardening background
  familiarity with function, operation, and maintenance of equipment considered appropriate for given site
  formal training in organic vegetable and fruit production
- **Knowledge of organic or Biodynamic gardening or farming methods:**
  philosophical commitment to organics
  awareness of resources available
  membership in state or regional organic farming association
- **Ease with people:**
  evidence of organizing skills
  evidence of teaching ability
  willingness to attend meetings with CSA core group
- **Personality traits useful for successful CSA farming:**
  well-organized
  observant
  determined and conscientious
  hard-working
  completes tasks
  self-starter
  able to share control
- **If a farming team:**
  are members cooperative?
  do they divide tasks and responsibility appropriately?
  do they have a clear decision-making process?
  are they respectful of one another?

the situation and agreed to share the high risks, it could work. But if the farmer was not able to get up to speed within a few years, the goodwill of the members would probably evaporate. In the brochure announcing the new CSA at Maysie's Farm in Glenmoore, Pennsylvania, Sam Cantrell explains his decision to give up field biology and settle down to farm the family land:

> We could "grow" houses, as most of the other farms in the area have done. Or we could again rent the fields out . . . Or we could accept the responsibility that comes with land ownership *and* ecological ethic and implement a management plan that sustainably utilizes the resources available. One of us could forego the advancement of his career and instead, invest his energy into the resurrection of the farm. That's the choice I've made . . . to have the farm become my career, to utilize this wonderful resource to fulfill my commitment to conservation and education.

Although Sam's promotional material sells shares as a contribution to the ecological preservation of land, he has made sure he also knows how to grow high-quality vegetables by getting two years of experience marketing to local stores before venturing into CSA.

FOR THE INCREASING NUMBERS of bright, energetic young people who want to commit their lives to sustainable living, CSA is one of the most promising paths. Many efforts currently in motion will make this easier: new internship and hands-on training programs; land-linking to connect would-be farmers with retiring ones; the Equity Trust Fund for CSA, which helps with planning and financing; the community food security projects introducing low-income city residents to organic food production and CSAs, and much more. Hopefully, in the near future, people who need a farmer will have many more skilled practitioners from whom to choose, and those who wish to farm will have many more ways to acquire the resources and skills they need.

THE LAND 5

Whenever there are in any country uncultivated land and unemployed poor, it is clear that the laws of property have been so far extended as to violate natural right. The earth is given as common stock for men to labor and live on . . . The small landowners are the most precious part of the State.

—THOMAS JEFFERSON, *WRITINGS*, V. XIX

WITHOUT A PIECE OF FERTILE GROUND, a CSA cannot exist. CSAs can acquire the use of land in a variety of ways, but for long-term, secure tenure of that land, there are many obstacles to overcome. An existing farm can convert all or part of its production to Community Supported Agriculture. An institution, such as a church, hospital, school, or food bank, can dedicate land it already holds, or purchase land for CSA use. A new farmer or consumer group can rent or purchase land for a CSA. Land trusts can play an important role in any of these variations. Each of these is a special case with specific benefits and difficulties.

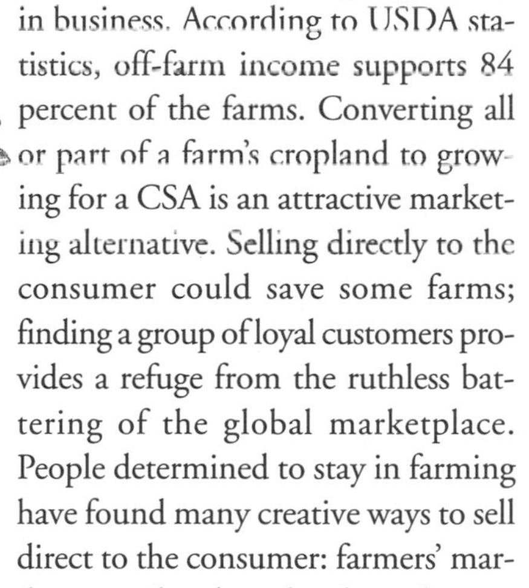

In his survey of CSAs, Tim Laird found that 59 percent of the CSAs are based on farms that have converted totally or in part from other forms of marketing; 76 percent of this group purchased or inherited the land and 17 percent received it as a gift or donation. The other 41 percent of CSAs were using land leased from private owners, land trusts, or for profit groups. In several cases, the rental payment was a share in the produce.

All over this country, family-size farms, ranging from two acres to one thousand acres, are struggling to stay in business. According to USDA statistics, off-farm income supports 84 percent of the farms. Converting all or part of a farm's cropland to growing for a CSA is an attractive marketing alternative. Selling directly to the consumer could save some farms; finding a group of loyal customers provides a refuge from the ruthless battering of the global marketplace. People determined to stay in farming have found many creative ways to sell direct to the consumer: farmers' markets, farm stands, or mail-order sales through catalogs or the Internet. In one sense, CSA is the latest wrinkle. And yet, as many farmers have discovered, the community-building aspect of a CSA can make it so much more than just a new form of marketing.

JOHN PETERSON GREW UP ON his family's farm in Caledonia, Illinois, tending dairy cows, beef cattle, chickens, and hogs, and growing field crops of corn, soybeans, and grains. He took over the 186-acre farm in 1969, when his father died. Over the next decade, he expanded production to 800 acres. Then disaster struck. Like many good farmers in the early 1980s, John was overwhelmed by debt, which forced him to sell most of the land, equipment, and livestock, leaving him with only 22 acres and the farmstead. After a few years of anger, despair, wandering, writing, acting, and offering counseling services to other distressed farmers, John returned to the farm, decided to grow vegetables organically, and gave birth to Angelic Organics. With three years of experience with wholesaling vegetables and farmers' markets, John and his then farming partner Kimberely Rector converted most of their land to CSA. By 1998, Angelic Organics was supplying eight hundred shares in and around the city of Chicago, renting 20 acres of neighboring land for $2,500 on a year-to-year basis, and concluding the purchase of a 38-acre piece of land to their north, which allows them to fallow one third to one half of the land each year. Rather than incurring further debt, the farm appealed to CSA members to become investors in a limited liability company that would purchase the land at $4,500 an acre, then lease it to Angelic Organics for fifteen years with the provision that the farm can purchase the land at fair market value at any time. Simple in concept, this approach turned into a legal labyrinth from which the farm emerged with an eighty-page legal document that cost an agonizing five months and $14,000.

Angelic Organics
Caledonia, Illinois

A HARD FREEZE AND LOW WHOLESALE PRICES for lemons and avocados drove Steve Moore to turn to CSA. After the Civil War, Moore's great-grandfather had homesteaded on the sixty-acre ranch in Carpinteria, California. Moore's father, a retired physician, inherited the farm in the 1960s. He hired a manager to oversee the production of lemons and avocados for the wholesale market. In 1981, he signed the farm over to his five children. That year, Steve writes, "was the worst year in history for the lemon and avocado markets. We just lost our shirts." Steve, with a Ph.D. in civil engineering and a masters degree in psychological counseling, found himself running the farm "working one hundred hours a week and making $500 a month." Gradually, he converted the farm to Biodynamic management. Then, in 1990, a big freeze destroyed the tree crops, a loss of $250,000. To save the farm, Steve turned to vegetables. "I knew," he later said, that "there was a community of people out there who wanted vegetables from me." In 1997, the Moore Ranch CSA included 220 families in the San Fernando and San Gabriel Valleys, Thousand Oaks, Santa Monica, and Malibu.

AFTER "DIDDLING" ALONG FOR SOME YEARS, Molly and Ted Bartlett of Silver Creek Farm in Hiram, Ohio, realized they needed to expand production in order to be able to afford to continue to farm. They spent several years in a frustrating attempt to coordinate a four-state truck route with other farmers. A contract for five acres of carrots fell through just before harvest time. They were able to sell most of the carrots through CSAs in the Washington, D.C., area. Then Molly met Robyn Van En, who asked, "Why don't you start doing it right at home?" That idea clicked, and in 1992, Silver Creek started a CSA with a goal of twenty-five shares. They targeted a mailing to friends and customers within one zip code and quickly recruited forty-seven members. Within a few years, membership swelled to 165, but they have since cut back to one hundred shares, which allows them to get to know all the members well. Silver Creek still runs a market at the farm twice a week, attends a farmers' market, and wholesales to store and restaurant accounts. They have added lamb, beef, chicken, egg, cut flowers, goat's-milk cheese, fiber, and "preserver" shares to fill more of the food needs of their CSA members.

ACQUIRING LAND

People who are thinking about starting a CSA often ask how many shares will support a farm and how much land will that require? The second part of the question is easier to answer than the first: Many farms are able to supply at least twenty shares per acre. The problem of scale is more complex, and closely linked to intensity of production. A biointensive grower with enough labor available can make a living on two acres. A more mechanized style of farming requires fewer helpers, but more land. For a single farmer to make enough money to live on, the minimum number of shares is around one hundred, if the CSA is the only market. The Gregsons of Island Meadow Farm in Washington State support themselves on the production of two acres of vegetables. They sell around thirty rather gourmet and modestly sized shares, and have other markets as well. In interviewing farms for the chapter on "CSAs That Quit," I found that a major reason they gave for quitting was inadequate income. Many of those CSAs were using a lot of hand labor, and had thirty shares or fewer. At the other end of the spectrum, several much more mechanized farms with four hundred shares and up seem to be making a decent living, and have funds to continue to capitalize their farms. But no one in his or her right mind would start a CSA with four hundred shares! That is a scale one has to grow into. Anything over two acres requires the use of mechanical equipment; the choice of the appropriate level of technology is a critical factor in determining whether you work yourself to death for a pittance, or run an efficient and at least modestly profitable farm enterprise.

A careful business plan is a basic requirement for obtaining any commercial loan. A list of CSA members who are committed to buying your farm's products should be proof that you have a market.

New farmers or consumer groups who want to start a CSA face the challenge of finding a suitable piece of land. Fortunately, several good books tell how to find land in the country, with detailed advice on what questions to ask and how to get through the legalities of purchase with the minimum of legal costs. *Finding and Buying Your Dream Home in the Country* by Les Scher, despite its silly title, is a helpful book. Vern Grubinger's new book, titled *Sustainable Vegetable Production on One to One Hundred Acres*, highly recommended for all new produce farmers, contains an excellent summary on finding land specifically for vegetables.

As Grubinger stresses, the quality of the soil is more important than the state of the buildings. Soil quality is crucial for organic production. While most soils can be improved by building up organic matter with compost and cover crops, contamination with heavy metals and persistant chemicals such as dioxin and PCBs cannot be eliminated. You should learn as much as you can about the history of the use of a piece of land, since testing is expensive. What was grown there? How was it produced? Was an apple orchard treated with arsenicals, DDT, or mercury? Is there a good source of water? How pure is the water supply? Do the neighbors use chemicals, air blast sprayers, or aerial spraying? Knowing specifically what to test for reduces the cost. Talking to neighboring farmers, and examining the soil maps at the county extension office or the Natural Resources Conservation Office will provide much of the information you need about the property you are considering and adjacent properties.

Residues of some pesticides are distressingly persistant. On August 15, 1996, the Connecticut State Department of Consumer Protection tested produce grown at the Holcomb Farm. The CSA managers were shocked when the results showed residues of Chlordane and DDE, a breakdown product of DDT. The pesticides were used on the tobacco grown at Holcomb Farm back in the early 1970s before they were banned. While the levels of the residues were low, at or below the Environmental Protection Agency "action level," the Hartford Food System, which oversees the Holcomb Farm, has enlisted the advice and help of a wide range of experts and agencies, sharing the information collected with the CSA members. Dr. Ted Simon, a toxicologist from the Center for Disease Control, volunteered to conduct a formal risk assessment. He came to the conclusion that "the lifetime risk of cancer from consuming about eight ounces per day of Holcomb Farm vegetables throughout the twenty-week growing season for thirty years would be about one in a million." According to John Cuddy, a bioremediation specialist in Wisconsin, the organic practices used on the farm seem to be the most effective means for stimulating the microbial activity that reduces chemical residues to harmless compounds. Similar residues were found on another organic farm in Connecticut even after thirty years of organic management.

CLEMENS KALISCHER

Community Supported Agriculture can help protect agricultural land such as this farm in Tyringham Valley, Massachusetts.

Some aspects of land acquisition are particular to CSA farms. The location of the land is important to consider. If your plan is to have pickup at the farm, you will need to be within a half-hour's drive of the population you hope to serve. CSAs that deliver to pickup sites in town can be farther away: the Decater's produce is trucked four-and-a-half hours from Live Power Community Farm to San Francisco. Several farmers have reported CSA members are harder to recruit in rural areas, though rural Iowans, even farm families, are joining new CSAs in that state. Urban and suburban dwellers are often more likely candidates.

Land purchase within half an hour of almost any city may be prohibitively expensive. Leasing is a possible alternative. If you find land you can lease, the owners need to understand that a CSA is not the typical farming arrangement. Agricultural lease agreements are usually made for conventional production systems where the farmer plows the ground, plants it, puts down his herbicides, and then comes back a few months later for harvest. CSA production involves daily attention and potentially large groups of people coming and going to the farm. A landowner who approached me about land he wanted to rent suddenly switched his tone from friendly to chilly when he learned I have two hundred helpers. Courtesy suggests that if the landowner lives near the CSA site, a day of rest from the traffic involved should probably be made part of the agreement.

Developing land for organic management is a long-term investment. The environmental stewardship this offers may appeal to some landowners. If they have any thoughts of selling the land within a few years, however, even the lowest rent would not make your investment in soil-building worthwhile, unless you are willing to take a chance as Bill Brammer is doing by farming land that is slated for development in San Diego (see p. 53). The best arrangement is a clearly written lease for a term of three to five years, giving the CSA the option of first refusal and applying the rent toward the eventual purchase of the land. The lease should also include consideration of any improvements the CSA makes on the property. The E. F. Schumacher Society has published a booklet entitled "A New Lease on Farmland," which suggests how such improvements can be valued so that both parties get a fair deal.

Where leasing is the best or only alternative, CSAs should consider asking for a three- or five-year rolling lease to ensure that they have adequate notice before being forced to move. Jennifer and John Bokaer-Smith have a five-year rolling lease with the Eco-Village at Ithaca, Inc. The section of the lease describing its term reads:

1. This Lease shall be a five-year rolling Lease.
2. The Term shall commence on May 1, 1997, and shall continue in full force and effect until May 1, 2002.
3. The Lease shall be automatically renewed for one additional year, to a maximum of five years, on May 1 of each year succeeding the commencement of this Lease.
4. Termination of the Lease shall be effected **only** by written notice given at least six months prior to the rolling renewal date. Such termination shall have no force and effect until five years from the date of notice.

USDA Farmers' Bulletin #2163, "Your Farm Lease Checklist," contains all the clauses you should think about including. The Pennsylvania State Extension's bulletin, "Guidelines for Renting Farm Real Estate in the Northeastern United States," has worksheets for calculating the rental fee. There are probably similar guides to agricultural rentals in other parts of the country.

THE FIRST YEAR AT INDIAN LINE FARM, Robyn leased the five acre garden site on the land she owned to the CSA for three years with the option to buy the site into an Agricultural Trust after the third year. Robyn designed this arrangement to preserve the acreage in perpetuity for the CSA, regardless of who was living in the farmhouse. Robyn understood the long-term needs of the CSA and believed deeply in its potential. Even with her positive attitude, the sale did not work out. Other landlords could not be expected to have her level of understanding, especially during the first few years.

IN 1988, A GROUP OF PEOPLE FROM the Albany area visited the E. F. Schumacher Society and heard about the Indian Line CSA. They decided they wanted to start one in New York. They wrote an article for the food co-op newsletter inviting people to a meeting to talk about finding land

and a farmer. Two landowners attended and both offered their land. Janet Britt, who grew up on a farm in central New York, had farmed in Colorado and Missouri and worked at the Indian Line CSA. She had never farmed on her own before, but she took the plunge to be their farmer.

Janet visited the two potential pieces of land and advised the group to select a five-acre piece on a farm in Schaghticoke. The owners were both doctors and agreed to a generously modest rent—a share in the produce. They also purchased a used tractor and leased it to Janet for $750 a year. Since they lived elsewhere, Janet acts as caretaker of their land and rents the farmhouse. The Hudson-Mohawk CSA has been running since 1988 with an average of one hundred shares per year. Until 1996, there was no written lease. Another farmer leased the hay fields and had a five-year lease that satisfied the local requirements to qualify for the farm rate on property taxes. When he dropped his lease, the owners offered Janet a five-year lease, so they could continue to enjoy the tax break. There is no written option to buy, but the owners have indicated that they are willing, though not anxious, to sell. They would like the land to remain a CSA. Janet reflects, "As I look back, I see what I should have done. I should have delineated money for a land fund in the CSA annual budget." Resolving the problem of more permanent tenure on the land is high on Janet's agenda.

TWO SEASONS OF GROWING on rented land and selling through the wholesale market in Boston convinced Tom Cronin and Chris Yoder to try direct sales. In 1994, they started Vanguarden CSA, growing the food on four pieces of rented land in the town of Dover, Massachusetts. Sue Andersen joined their team for two years, from 1995 to 1997. By 1996, they had reduced wholesaling to 10 percent of production and increased shares to 193. Using rented land is a limiting factor, discouraging them from investing in perennials, such as tree fruits. It is also expensive: their annual budget includes $8,850 for rent. During the winter of 1996–97, the Vanguardeners decided to begin a search for a permanent piece of land. With some of their members, they formed a land search committee, which surveyed the members by phone to find out what skills or resources they could offer. The committee also contacted state, regional, and town conservation groups to try to locate potential sites. Committee members with professional fundraising and financial skills are currently working on a funding plan.

THE STONY BROOK-MILLSTONE Watershed Association funded the establishment of a farm on a piece of its land in Pennington, New Jersey, in the early 1980s, and hired a farmer. The association quickly learned that a two-acre vegetable garden could not pay the farmer's $25,000 salary. By the time Jim Kinsel took the job, the Watershed Association was still providing free use of the land, a two-bedroom house, equipment, and volunteer labor, but expected him to earn his own salary from the farm business. Since Jim has turned the farm into a successful operation, the association has decided to maximize its return on the property and has required substantial lease payments for use of the farm. In 1997, Jim agreed to a three-year lease with a straight landlord/tenant relationship. The annual rent is $25 an acre, $900 a month for the apartment Jim lives in, and $3,000 for the farm buildings for the first year, rising to $7,000 by the third year, calculated as 75 percent of the market

Vanguarden CSA started on rented land.

value. Jim says, "I still count my blessings in renting from an organization. There's a kind of check and balance on the board; no one individual can throw me off."

IN THE RAPIDLY EXPANDING SPRAWL of San Diego where land is priced for developers, not farmers, members have played a critical role in helping Be Wise Ranch get access to land. Bill Brammer owns 20 acres and leases 300. A senior planner and other influential members of his CSA helped him obtain a lease that will run from fifteen to twenty years on 150 acres that belong to the City of San Diego. The rest of his leases are short-term, on land in line for development. One parcel, the property of a professional golfers' association, is currently zoned for 10 acres per house. The association wants to increase the density to four or five houses per acre, but voter approval is needed to change the zoning. Bill is betting the voters will say no, and that the process will drag on for years while he continues to farm the land. His members may be able to help him persuade the zoning board to institute a transfer of development rights program to keep more of the land open. In 1998, Bill had to relinquish a piece of land he had farmed for twenty years, forcing him to dismantle and rebuild his packing shed. The next crop on those acres will be houses.

According to American Farmland Trust, 4.3 million acres of prime and unique farmland were lost to development and suburban sprawl between 1982 and 1992, nearly 50 acres every hour.

While going into debt is not a good idea, borrowing money from a bank may be the only way for some farms to obtain funds to purchase farmland. Many rural communities have banks that specialize in agricultural lending. Local farmers or the Cooperative Extension should be able to tell you which banks to approach. A careful business plan is a basic requirement for any commercial loan. A list of CSA members who are committed to buying your farm products should be proof that you have a market. Resources are available to help you develop a business plan. In some states, the Extension has economic specialists who will help you. The Service Corps of Retired Executives (SCORE), a national network of volunteer business executives and professionals, offers technical and managerial counseling to small businesses, such as farms. (For information on your local chapter, contact the national SCORE office at 800-634-0245.) An advanced course in Holistic Resource Management could help you create a plan that is both convincing to a lender and truly useful for your farm's decision-making process. The Federal Farm Service Agency (FSA), serves as the lender of last resort, if commercial banks turn you down. The FSA is supposed to reserve some of its funds for beginning farmer loans. (A beginning farmer is someone with less than ten years experience.) In addition, a few states, such as Iowa and Nebraska, have "aggie bond" programs to encourage loans to new farmers. Dan Looker's *Farmers for the Future* documents creative ways in which the truly determined can get into farming. The most radical (or perhaps the most conservative) aspect of Rose Valley Farm was our total avoidance of borrowing. If we had the money to buy something, we paid in cash; if not, we waited. The need to make mortgage or other debt payments can add an unbearable level of pressure to running a farm and have a profound effect on all management decisions. Avoid it if you can.

PROTECTING FARMLAND

Securing land for farming for the long term is a problem shared by farms of all scales and forms of ownership. According to American Farmland Trust, 4.3 million acres of prime and unique farmland were lost to development and suburban sprawl between 1982 and 1992, nearly 50 acres every hour. Even CSA farmers who own their land outright with no mortgage or debt are seeking ways to ensure that the land remains in farming through future generations. Organic and Biodynamic farmers want to keep the land they are working free of chemical agriculture. The legal tools available include conservation easements, land trusts, and the purchase or transfer of development rights, but none of these offers a quick and easy formula.

Chuck Matthei, president of Equity Trust and a pioneer in preserving lands from the inflating pressures of the real estate market, has written in "Gaining Ground: How CSAs Can Acquire, Hold and Pass On Land" (available from Equity Trust, see CSA Resources) that the key to finding the resources to finance farmland protection "lies in distinguishing the essential personal interests in farm properties from the inherent public interests. Defining and protecting the public interests legitimizes the application of charitable and public funds to a land purchase, and thereby assures affordable access and full opportunity to the farmer."

In "This Land Shall Be Forever Stewarded," Jered Lawson delineates the private and public interests:

The private interests in farming may include:

1. secure permanent tenure for the farmer;
2. equity in the property; which is affordable out of farming income, transferable to heirs, and recoverable through sale if necessary;
3. freedom and flexibility in managing the farm.

The community interests in farmland may include:

1. an ongoing source of quality food (maintaining the land in active production);
2. care taken for the long term fertility of the farm (preventing agriculturally generated pollution and environmental deterioration);
3. ensuring access and affordability for successive farming, and preventing the deleterious effects of absentee ownership. (Lawson, pp. 2–3)

Particularly in parts of the Northeast, where farming has all but disappeared, a large number of people feel strongly about the need to protect the farmland that remains. Besides the landscape value of the farms, studies focusing on cost of community services have shown conclusively that communities save money by investing in farmland and limiting development. State and local governments have established Purchase of Development Rights (PDR) and Transfer of Development Rights (TDR) programs. Although these programs operate differently from state to state, basically, PDR programs assess farms at their value as farmland and at their value as real estate, and pay the farmer the difference. In exchange, the state or town receives an easement, which is written into the deed on the property, eliminating all future development. In most cases, the effect of PDR is to reduce the value of protected properties, so that they remain affordable as farmland. PDR does not protect the activity of farming, nor does it ensure any particular method of farming. The Massachusetts PDR program changed its easement to require continuing agricultural use. The Food Bank CSA in Massachusetts was able to acquire land at a reasonable price because the Commonwealth had purchased the development rights.

TDR is more complex: Through zoning, communities establish "sending" and "receiving" zones. To develop more densely in a receiving zone, a builder must purchase development rights from a sending zone. The cost to taxpayers is lower for TDR, but the administrative and legal complexities limit its use.

Vermont has adopted the most aggressive approach to saving farmland: the Housing and Conservation Board holds a statutory right-of-refusal on any farm that has received property tax consideration or other state subsidies, before it can be sold and removed from production. The Vermont board also provides financial support to local land trusts through a fund capitalized by state tax money (Matthei, "Gaining Ground").

Land trusts provide a nongovernmental alternative. There are two kinds of land trusts: conservation land trusts and community land trusts. Usually, both are locally based, member-controlled, nonprofit corporations. The same legal and financial tools are available to both kinds of trusts. Most conservation trusts are dedicated to preserving open space, though about 10 percent of them have an active interest in farm and forest lands. Conservation trusts usually hold an easement on pieces of land, leaving the fee interest or title in the name of the farmer.

Community land trusts are largely urban, and are devoted to providing the essential benefits of ownership to low-income people who have been excluded from the real estate market. A few, such as the Pioneer Valley Community Land Trust in Massachusetts, have farm holdings. Community land trusts usually retain title to the land, which they lease to residents. *The Community Land Trust Handbook*, by the Institute for Community Economics, provides an introduction to this form of land trust, with examples from around the country and guidelines for organization. The leases are often for ninety-nine years and inheritable; lessees own the buildings and other improvements. The Pioneer Valley Community Land Trust requires ecological management of the land it holds.

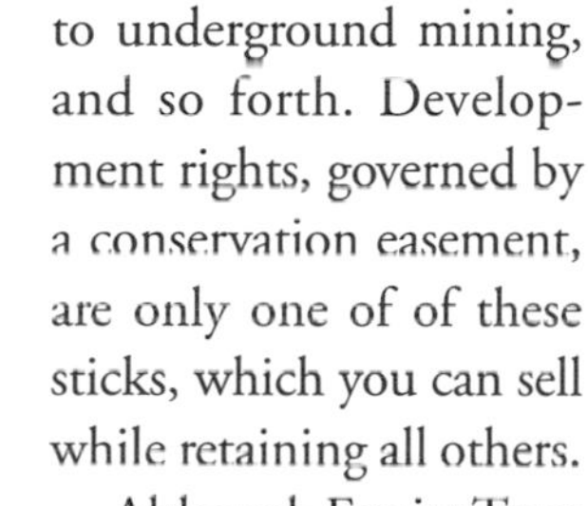

Live Power Community Farm
Covelo, California

In finding a way to protect your farmland, the choice between a conservation trust and a community land trust may depend more on what organizations are available locally. According to Chuck Matthei, the "difference between the individual holding title and the land trust doing so may seem significant, but in fact, it may not be." More important are the details of the legal agreement, whether lease or easement.

In *The Conservation Easement Handbook* Janet Diehl defines a conservation easement as "a legal agreement a property owner makes to restrict the type and amount of development that may take place on his or her property."

> Each easement's restrictions are tailored to the particular property and to the interests of the individual owner. . . .The owner and the prospective easement holder identify the rights and restrictions on use that are necessary to protect the property—what can and cannot be done to it. The owner then conveys the right to enforce those restrictions to a qualified recipient, such as a public agency, a land trust, or a historic preservation organization." (p. 5)

The *Handbook* is a guide to understanding and writing conservation easements, providing examples, as well as suggestions for monitoring and information on tax law. I have heard Chuck Matthei and others compare property rights to a bundle of sticks, which include the rights to ownership, to access to the land, to underground mining, and so forth. Development rights, governed by a conservation easement, are only one of of these sticks, which you can sell while retaining all others.

Although Equity Trust is willing to hold conservation easements on CSA farms, Matthei recommends that farmers develop a relationship with a local land trust. Defining the terms of the easement on a farm and the oversight this will require from the trust may entail a long process of mutual education. Nevertheless, this will probably be easier than creating a new organization for the single purpose of protecting your farm. In beginning to explore these issues, I inquired of American Farmland Trust whether they would be willing to hold an easement on Rose Valley Farm. My partner David and I owned the land outright, but we were hoping to guarantee ecological management after we are gone. Jerry Cosgrove, who runs AFT's New York office, told me AFT would accept a conservation easement, but does not have the capacity to ensure that the management of the farm remains organic. He recommended I find a local organization.

THROUGH IMPRESSIVE HELP from their communities, Live Power Community Farm and Fairview Gardens, both in California, and Common Harvest CSA, in Wisconsin, have made the transition from leasehold to protection in perpetuity. After farming on borrowed land for over two decades, Gloria and Stephen Decater of Live Power Community Farm faced the possibility that Richard Wilson, the owner, might need to sell the land, and that they would not be able to afford to buy it at its market price. Four seasons of weekly vegetables had attracted a group of loyal farm members in Covelo and San Francisco. The Decaters turned to them for help and advice. Passions in the group ran high: there was consensus that the Decaters were the stewards everyone wanted for the land, and that the Decaters deserved greater security. But the group divided over how to accomplish this; some people were in favor of private ownership, while others preferred a nonprofit arrangement.

The Decaters themselves were more concerned about the responsibilities of stewardship than private property rights. Stephen commented:

> When we contemplated private ownership, we had two areas of concern. One was that to acquire complete private ownership would entail placing ourselves in a major debt for an extended period, even if we could find a lender. We were opposed to the concept of entering into debt, and did not have a credit history, having always operated on a cash basis. The other concern was that we saw one's relationship to the land as having moral dimensions. We saw land ownership not as acquiring strictly a commodity only to be treated as we please, but also as an office bearing certain inherent responsibilities requiring wise (just) usage and stewardship. We saw land as a resource that also belonged to future generations, a resource, which we had no personal right to appropriate or damage regardless of whether such a treatment was legal by current law or not. We wanted to be sure that the form of legal ownership we entered into would reflect this moral imperative. (Jered Lawson, *This Land Shall Be Forever Stewarded*, p. 7)

Four years of meetings and fundraising, of encouraging advances and sudden dead ends, led to a successful combination of public and private ownership. The Decaters were able to purchase their 40 acres of farmland at its value based on its potential for earning agricultural income typical of the area. This resulted in a private investment of $69,000 ($20,000 for land and $49,000 for buildings) for the Decaters, while the CSA members raised $81,000 to purchase the development rights and all other non-agricultural use rights of the property. They could not come to terms with a local land trust, so Equity Trust in Connecticut will hold these rights in a conservation easement on the property. The easement is a remarkably detailed document that tries to foresee every eventuality. The terms of the easement place permanent responsibility on Equity Trust (or its successor organization) for oversight of the management of the land. The "Purpose" clause of the easement reads:

> It is the purpose of this Easement and Option to assure that the Property will be forever biodynamically or organically farmed and, as to that portion of the Property not farmed, retained predominantly in its natural, ecological, and open condition and to prohibit any use of the Property that will impair, degrade, or interfere with the conservation values of the Property. Without limiting the general purpose of this Easement and Option, the following specific purposes are intended for this Easement and Option:

(a) To conserve and protect the Property's agricultural, natural and ecological value and prevent environmental pollution and degradation due to the use of agrichemicals;
(b) To prevent the conversion of agricultural land to urban and nonagricultural use; and
(c) To assure that the Property will be farmed, with organic or biodynamic methods, for the production of plant and animal products for commercial or charitable purposes.

To enable Equity Trust to oversee these purposes, the easement requires that the Decaters or their successors provide reports on any intended changes at the farm, and documentation of the annual organic or Biodynamic certification of the farming methods. Since the easement stipulates that the landowner must make at least half of his or her income from farming, the Decaters must also submit a copy of their income tax return. Equity Trust also retains the right to inspect the property with forty-eight hours notice. Should the Decaters or their successors violate the easement, Equity trust has the right to demand corrections and reparation. Should the Decaters stop farming, or try to sell the land for more than its agricultural value, Equity Trust has the first right to purchase the land.

The Decaters themselves participated in the creation of this easement, which ensures that at any future sale of the property, the value of the land will be based on its agricultural earning potential so that a farmer will be able to finance it from farm income. It is not some form that a state office imposes upon them or any other farmer. Every farm entering into a similar arrangement has the freedom to choose the terms of a conservation easement to satisfy its own needs and desires for the future. As Stephen puts it, the easement "can be the vehicle for partnership between the private individual and a nongovernmental community-public representative." The Decaters' easement creates permanently affordable farmland, accessible on the basis of commitment to stewardship, rather than accumulated capital.

Other farmers have forged connections between community members and stewardship organizations to protect their farmland.

IN THE MINNEAPOLIS-ST. PAUL AREA, Dan Guenthner and Margaret Pennings have been running Common Harvest CSA on rented land since 1989. To purchase land of their own, they appealed to the 170 members of their CSA for help with either loans or donations. Forty percent of the members responded with amounts ranging from ten to several thousand dollars. Members loaned $12,000 in the form of prepaid shares for the next five years. The $28,000 in donations went to the Wisconsin Farmland Conservancy to purchase the development rights to the farm. "When we were meeting with Tom [Quinn, director of the Conservancy] to work out the details of the easement, we realized that this document amounts to a public declaration of what we intend to do with the land," Dan observed. "Other than getting married and having children, this is the most important commitment we have ever made. We are wedding ourselves, on all sorts of levels—spiritual, emotional, social—to this place" (*Growing for Market*, Feb. 1998, p. 9).

TO ENSURE THE FUTURE of Fairview Gardens in Goleta, California, Michael Ableman chose to give up any prospect of personal rights to the property and instead to form a nonprofit corporation. Ableman farmed the twelve-acre piece, first as an employee, then as an independent contractor, and finally under a variety of leases. Meanwhile, the suburbs closed in around the hundred-year-old farm, the one lingering residue of what was once a large rancho typical of the predominent land use in the fertile area near Santa Barbara. Taking the nonprofit route was a difficult decision for Ableman: "This was my life, sweat, and blood for seventeen years, my school for myself and my family. I will have no monetary ownership."

In 1992, after the death of one of the owners, Ableman realized that the time had come to take energetic measures to protect the property. The market value of the twelve acres was over $1 million, an unlikely sum to raise to save a tiny family farm. Ableman's strategy has been to stress the public values inherent in Fairview Gardens. He set out a mission that encompasses both food production and public education. The mission statement promises to:

- Preserve the agricultural heritage of this hundred-year-old farm;
- Nurture the human spirit through educational programs and public activities at the farm;
- Demonstrate the economic viability of sustainable agricultural methods for small farm operations;
- Research and interpret the connections between food, land stewardship, and community well-being;
- Provide the local community with fresh, chemical-free fruits and vegetables.

Ableman approached the Santa Barbara Land Trust to hold and monitor a conservation easement on Fairview Gardens. As the Decaters found with their local land trust, an easement that specifies the continuation of farming, and insists on organic methods to boot, presents a challenge. The Santa Barbara Land Trust hesitated about making the commitment, but timely intervention by Chuck Matthei of Equity Trust helped reassure them that overseeing a farm is within their mission. The Fairview Gardens easement presents an additional complication since it also requires that the educational work of the nonprofit be maintained in perpetuity. This is new territory for conservation land trusts.

Impressive local support, persistance, savvy organizing, and sheer hard work culminated in the October 1997 celebration of the purchase of Fairview Gardens. Jean Schuyler, a community member and generous contributor, headed up the fundraising campaign to save Fairview. Public funds for preserving farmland amounting to $146,000 combined with private contributions to cover the $750,000 purchase price for the land. Besides this money, the landowners will also realize substantial tax savings from the transaction. Various costs for the fund-raising campaign, legal fees, and the like amounted to another $45,000. Fairview will also need to raise additional funds to provide an endowment for its educational programs and to support the Santa Barbara Land Trust in monitoring this unique conservation easement.

Michael Ableman has concluded from this successful, but grueling experience that he feels an obligation to help others who wish to protect their farmland. "To raise charitable funds for a private business," he observes, "is pretty near impossible. If the farm extends into community action, such as an educational component that brings children to the farm, that changes the dynamic." Without the help of someone like Jean Schuyler, a person who could appeal to her peers for support, a campaign such as Fairview's would not have been possible.

◆

IF OUR AGRICULTURE is to become a support for local communities, we will have to become much more skillful at protecting farmland. As Chuck Matthei points out, "Although we define the word 'equity' both as a financial interest in property and as a moral principle of fairness, all too often it seems that we have forgotten the necessary relationship between the two" ("Gaining Ground," p. 9). It may be that we have forgotten, and perhaps this connection is one of the many lessons we need to relearn in building the community aspect of CSA.

— Part III —

Getting Organized

NURTURING A SOLID CORE GROUP 6

Somewhere, there are people to whom we can speak
with passion without having the words catch in our throats.
Somewhere a circle of hands will open to receive us, eyes
will light up as we enter, voices will celebrate with us
whenever we come into our own power.
Community means strength that joins our strength to
do the work that needs to be done.
Arms to hold us when we falter.
A circle of healings. A circle of of friends.
Someplace where we can be free.

—STARHAWK

FOR CSAs TO BE MORE than just another direct marketing scheme, the growers and the eaters of the food need to work together to build an institution they can share. In a consumer-initiated or institution-initiated CSA, the hired growers must feel they are more than temporary employees who serve at the will of a board that may not understand a great deal about food production. If a farm recruits people to subscribe to its produce, the growers must relinquish some of their autonomy to make decisions for their farm business. By tradition, we have given the name "core group" to the grower-member council in which we come together to run our CSAs.

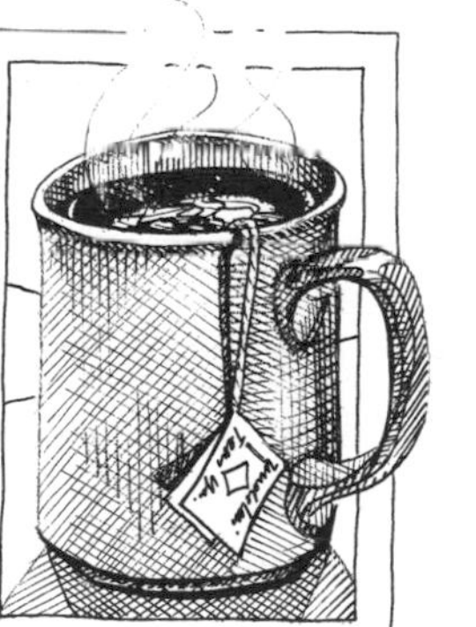

To be sure, not every CSA has a core group. Angelic Organics only organized a core group after the CSA was firmly established, when the farmers had some extra energy. A few farmers, like David Inglis at Mahaiwe Harvest, insist that they do not need administrative or farming help from their members. David says that he and his family have the farm work, bookkeeping, and associated efforts firmly under control. The CSA members are still, according to David, 100 percent more active in relation to the farm than to any store where they purchase food. Many other farmers would envy David's self-assurance and ability to cope. In fact, one of the main reasons farmers have given for why they quit doing CSA has been their failure to find members who would help them.

A CSA is in essence a member-farmer cooperative, whoever initiates it and whatever legal form it takes. As a cooperative, a CSA is a hybrid enterprise blending worker control and customer control. No universal formula or recipe exists for creating a CSA, but organizers can tap into a rich tradition to find examples from which to learn. Its roots go back to the founding of the Rochdale Cooperative in England in 1844. Rochdale intended to offer members groceries, housing, clothing, manufacturing, jobs, and "a self-supporting home colony of united interests."

The first step in building a core group is to find a few people who share a commitment to the project. Our CSA, Genesee Valley Organic (GVOCSA), was fortunate in its origins. We began as a mutually supportive effort of the Politics of Food (POF), an organization devoted to food issues in Rochester, New York, and David and me, the farmers. We quickly attracted other people with years of organizational experience in the peace movement, the Catholic Worker, churches, home schooling, environmental groups, and business, as well as some newcomers to community work. This group could come up with a better solution to any organizational problem than David and I ever could on our own.

The second step, essential to involving committed people, is designating the group's powers and giving it real responsibility. In the GVOCSA, we agreed from the beginning that David and I would make all production decisions about the farm: how to grow the food; what seed, fertilizers, and other materials to use; timing of operations; what to harvest when; and what equipment to purchase. The core group would not interfere in the farming in any way, beyond their initial decision to make a commitment to Rose Valley Farm. Our organic certification was an important factor in this choice. While we, the farmers, decide what is ready to pick each week, the members give guidance by filling out annual veggie questionnaires and making suggestions of favorite crops. Although some CSA core groups discuss the annual budget, I have not heard of a CSA in which the members actually run the farm.

Resources for Organizing Cooperatives

If you are inexperienced, running meetings or creating a core or board may seem daunting. If you decide you want to learn, the following resources can help you improve your organizational skills.

- A basic reference is *Welty's Book of Procedures for Meetings, Boards, Committees and Officers*, available for $9.95 from Caroline House Publishers, Inc., 920 West Industrial Drive, Aurora, IL 60506; (312) 897-2050.
- New Society Publishers (4527 Springfield Ave., Philadelphia, PA 19143) offers resource books on group process and decision-making, such as *Manual for Group Facilitators, Resource Manual for a Living Revolution,* and *Democracy in Small Groups.*
- In the writings of E. F. Schumacher (e.g., *Small is Beautiful),* Richard Borsodi, George Benello, Hazel Henderson, and Barbara Brandt, you can delve into the theory of small-scale economics, and cooperative enterprises.
- The E. F. Schumacher Society (Box 76A, RD3, Great Barrington, MA 01230) has an entire library devoted to decentrism, and the Center for Economic Democracy (P.O. Box 64, Olympia, WA 98507) is compiling a collection of basic co-op documents.
- Food Not Bombs Publishing (295 Forest Ave., #314, Portland, ME 04101) carries C. T. Butler's books, *On Conflict and Consensus,* and *Food Not Bombs: How to Feed the Hungry and Build Community,* as well as other books.
- The National Cooperative Business Association (1401 New York Ave., NW, #510, Washington, D.C. 20006) provides advice and legal support.
- The USDA Agricultural Cooperative Service (USDA-ACS, P.O. Box 96576, Washington, D.C. 20090-6576) provides support services to beginning and existing agricultural cooperatives, and publishes a monthly magazine, *Farmer Cooperatives,* which is free to qualifying organizations.

The GVOCSA core group is empowered to make decisions about everything that happens after the food leaves the farm. With the agreement of the other members, the core has established the basic values of the organization: an emphasis on ecologically grown food from local farms, sharing the risks of crop production with the farmers, maximum involvement of children, and a striving for social justice. From setting the price for a share, to deciding what packaging to use or avoid, to arranging the schedule for work at the farm and distribution sites, to controlling the bank account, the core is also in charge of the day-to-day details. The core decides what other products to offer the members, such as organic strawberries, maple syrup, wine, cheese, and low-spray apples, and sets the quality standards. Four or five years ago, I suggested we offer nonworking shares at a higher price. The core group voted this down.

At about that same time, a quiet coup took place within the core. For the first years of the GVOCSA's existence, I acted as chair and facilitator of the monthly core meetings. After thirty-five or so years of experience with organizations, I have gotten fairly good at moving a group cheerfully through a long agenda, so people are often willing to let me do this job. In my head, the core members were doing a remarkable amount of work for my farm, but it was *my* farm. With one of those funny ironies of life, in the very same month that I brought the chapter on meeting facilitation from *The Resource Manual for a Living Revolution* to a NOFA Governing Council meeting, Dennis Lehmann, GVOCSA treasurer, distributed that same chapter at a core meeting! In his gentle, but firm way, he suggested we take turns facilitating the meetings, set the agenda as a group, and allocate the amount of time for each item. The core accepted his proposal in a speedy consensus. It may have been my farm, but it had become *our* CSA.

AT A WORKSHOP on "Member Events and Involvement," Elizabeth Smith of Caretaker Farm in Williamstown, Massachusetts, expressed a sentiment I felt and have heard from a lot of farmers: "Since we had an established farm, it's been hard for us to step out of the picture and let the members take over. We almost feel guilty that we are asking all these people to take on these roles." The Caretaker Farm "Handbook for Members" outlines their complex organizational structure:

The Steering Committee:
This committee is composed of the farmers and those who wish to shape the overall direction and organization of the farm. . . . Five sub-groups were created by the Steering Committee to strengthen the community and sustain it for the future.

Farming: This group is made up of Elizabeth and Sam and all the apprentices. They are responsible for the day-to-day farming decisions.

The Farm Economy: The Treasurer and the Steering Committee create the annual budget and monitor income and expenses.

Community Support: Responsibilities include recruiting distribution helpers and work support for the farm, planning our various festivals and celebrations, producing the newsletter and this Membership Handbook, running fundraisers as needed, and improving communication among the members.

Organizing: Responsibilities include putting the Annual Proposal together, keeping agendas and the minutes for the Steering Committee, and organizing the CSA's structure.

Future/Long-Range Planning: This group is working on our mission statement, exploring associative relationships with other CSAs and compatible organizations, and developing community outreach programs.

WHEN I ASKED Molly Bartlett of Silver Creek Farm in Ohio what was the hardest decision she ever had to make about her CSA, she answered that it was the decision to let some of the members become a working core. Her husband Ted agreed. Letting go of some of the responsibility for the farm has been very hard. Yet Silver Creek today has one of the most active cores of any CSA I have encountered. Two members coordinate the farm work for each of the three weekly work sessions while Ted and Molly "hover" nearby. A chief coordinator from among the members oversees the work of the other core workers. The core is so well trained that, when Ted had a heart attack, they were able to keep the farm rolling along.

Besides the conscious agreement the GVOCSA core group made in dividing up areas of responsibility, we have an unspoken understanding. We do not have to agree on everything. We have no CSA party line. We do not endorse political candidates, we have no policy on abortion, gun control, or gay rights. We *have* taken positions in favor of rBGH labeling on dairy products, and against food irradiation, and our newsletter provides information about NOFA, the New York Sustainable Agriculture Working Group, and the Campaign for Sustainable Agriculture. So far, we have not had anyone infiltrate the group with a hidden agenda. Were that to happen, a more explicit agreement on where we agree to agree might become necessary.

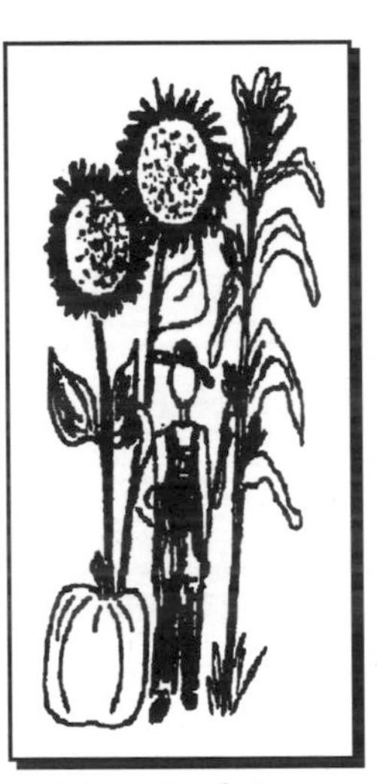

Silver Creek Farm
Hiram, Ohio

Every CSA has to decide how many people to employ, what jobs the employees should cover, and what work is to be done by volunteers. After two years of operation with twenty-nine and then forty-five shares, the farm suggested the GVOCSA grow to one hundred shares. At the end-of-season dinner meeting, we presented the members with two alternatives: the CSA could choose either to hire someone to do administration or expand our system of volunteers. The group voted overwhelmingly to do the work themselves, and to distribute the jobs as widely as possible to prevent burn-out. A 1991 winter letter appealed to the members to volunteer:

> We can build a strong community, a responsive organization, just as farmers can. The GVOCSA will only be as good as the effort we put into it. Each one of us *is* the organization, *is* the community. If we distribute the work, then it rests happily and securely on everyone's shoulders. Please consider joining the Core Group as a scheduler, process person, outreach, distribution, or children's committee member. In this way, we can build a CSA as sustaining as the soil in which the food is planted.

Our core group for 1997 consisted of twenty-two members: two farm representatives, two process people, two treasurers, a scheduler for each of the three weekly work days, nine distribution coordinators with a person to coordinate them, two outreach people, a social director, and the newsletter editor. The process people handle telephone inquiries, incoming mail, outgoing mailings, and generally oversee communications among the members. In 1995, we added the social director's position to develop ways of strengthening the feeling of community among the members. So far the job list includes setting up picnics for members, arranging for pickup parties once a month for each of the distribution days, helping organize the end-of-season dinner, and creating a members' directory so that everyone will have basic information about everyone else. All twenty-two members do not make it to all of our monthly meetings, but attendance rarely dips below twelve. In 1997–98, we reorganized the core jobs to reflect the changing realities of the CSA while we purchased produce from other farms as we moved from Rose Valley to Crowfield.

Clear work responsibilities are a necessity. We have a job description for every function in our core.

CSA Structure and Jobs

Farmers

develop farm budget to present to core
prepared field plans
soil preparation
seed selection
planting
cultivation
harvesting
machine work
maintenance and repair of tools and buildings
purchase farm supplies (fertilizers, pesticides, seed, fuels, packaging)
pay farm insurance and taxes
capitalization and financing
bookkeeping
- financial records
- production records
- certification

educate sharers about farming
oversee sharers' farm work

Sharers

decide on legal structure
decide on commodities to purchase jointly
agree on group values (e.g., organic, local, include low income, social diversity, involve children)
select core group
participate in CSA
- pay on time
- do farm work
- do distribution work
- help recruit members
- pick up and enjoy food share
- end-of-season survey to evaluate project

Note: These jobs can be combined in any number of ways depending on talents and energies of farmers and sharers.

Core Group as a Whole

selects farmers
selects land
selects crops desired
sets fee policy
sets payment schedule
determines work policy (amount of member participation)
sets distribution place(s)
sets distribution procedure
monitors progress of project

Job Assignments

treasurer
- collect fees
- pay farmers
- keep books
- maintain bank account

communications
- maintain newsletter and notices
- answer queries (calls and mail)
- post office box
- membership list
- educational materials
- annual contract

outreach
- recruit new members
- publicity
- links with other CSAs, farms, co-ops, other groups
- raise scholarship money

schedulers of member participation
- farm work distribution

distribution organizers
social director
- organize group activities (picnics, dinners, celebrations)

organize bulk orders from other farms
oversee involvement of children
- create play area on farm
- set guidelines for parents and clear rules for children

CLEMENS KALISCHER

Core group meeting at Caretaker Farm.

Despite our efforts to spread the work evenly, for each of the ten years of our existence, a few people have done a disproportionate amount of the work. Remarkably, few people have burned out, since each year the extra burden has been taken up by different people. Our universal work requirement and the forty miles separating our farm from the sharers has led to a structure that is more elaborate than farms with a closer location might require. But we are a living demonstration that distance from customers need not rule out member participation.

IN SITUATIONS WHERE THE FARM is located even farther from the members, CSAs really depend upon their city core groups to handle distribution. Roxbury Farm in Claverack, New York, has a separate core group for each of its three target areas. Each core determines its own budget for administrative and distribution expenses, which include the cost of free shares to members who take on site coordination. These cores organize distribution, oversee the weekly work shifts, recruit new members, handle fee collection and bookkeeping, and help plan farm festivals. Twice a year, the three cores meet together with the farm crew to discuss the annual budget and the overall direction of the farm.

MOST OF THE CORE MEMBERS for Live Power Community Farm live in San Francisco, a four and a half hour drive from the farm. This core acts as a long-range planning council and took the major responsibility for raising the money to purchase the development rights to the farmland, over $89,000.

(See chapter 5 for more details.) A section of the core is charged with cluster organization, shaping the members into convenient neighborhood groups for food pickup, and scheduling their work. Two core members make sure that there are good communications between the members and the distant farm.

IN THE FIRST ISSUE of the Angelic Organics core group newsletter, "A Farm Forever," Tom Spaulding explained that the core is made up of volunteers "so inspired by this particular community-supported Farm, that we dedicate our time, energy, wisdom, and resources to help the Farm fulfill its mission. The more we do as shareholders to help with non-Farm tasks, the more our hearty and visionary farmers can concentrate on producing the delicious, colorful, nutritious, aromatic, and simply splendid veggies and fruits that we have come to depend on from June to November." Like the Live Power core, the Angelic Organics group has taken on strategic planning and financial development, facilitating connections with the farm, and assisting the farm crew with city distribution. They also want to work towards community food security by arranging for leftover shares to go to a food bank and seeking funding to subsidize share fees for low-income families.

The GVOCSA core group makes decisions by consensus. At meetings, we discuss each new issue. If we cannot come to an agreement right away, we often table the question. By the time it comes up again (and some things never do—they either self-solve or self-destruct), agreement usually comes quickly. We keep meetings short—two hours a month, including refreshments and socializing. Although we must deal with an unending flow of details even after nine years, we keep the tone light, reserving our competitive drive for inventing the worst puns. While some boards recruit members based on particular skills, we tap members who demonstrate a good sense of humor. The stability of our membership is a testament to the satisfactions of being part of a group that functions well.

IN BUILDING OUR OWN LITTLE INSTITUTIONS, like the CSAs, we are at the same time transforming ourselves into people who can listen to one another and take cooperative action. Changing isolated farm enterprises into community supported farms cannot happen overnight. We have all kinds of inertia, internal and economic, to overcome, and a lot of new skills to learn. But our creativity is unlimited when we work together in a atmosphere of mutual respect. This next decade is a critical time for farming both in this country and the rest of the world. A local food system based on direct links between farmers and eaters is the opposite of the global market touted by our government, that benefits only the Cargills, Nestles, and Philip Morrises. Each new CSA is another little piece of liberated territory, and a step toward the sustainable world that is our only possible future.

LABOR 7

To allow everyone involved in organic production and processing a quality of life which meets their basic needs and allows an adequate return and satisfaction from their work, including a safe working environment.

—IFOAM Basic Standards

There are never too many hands to do all the work on a small farm and rarely enough dollars to pay for all the work that needs to be done. According to USDA statistics, off-farm earnings help support 84 percent of the farms in this country. The unpaid labor of the farm family is essential. When the workload swells beyond what the family can manage, farms must shoulder the expense of hiring workers or take on trainees. CSA farms are no exception, but they have the exceptional opportunity to get help from a new source—the members.

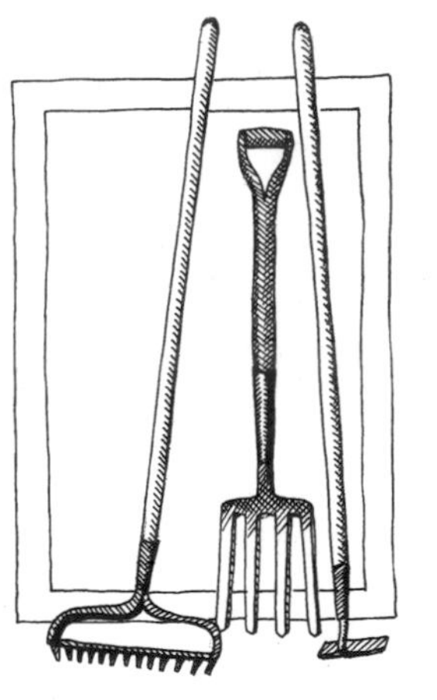

While CSA members and potential members, when polled, claim that they do not have time to participate, what people *say* does not always coincide with what they are willing to *do.* Everything depends on how you ask the question. If a survey asks, "do you want to work on the farm or take a job in the core group?" most people respond with a self-defensive "No!" At least two other approaches work better. If, from the beginning, farmers and members discuss how to keep share prices as low as possible, participation is an obvious way to reduce cash payments for the food. No one refers to the work as "volunteering." If the CSA then organizes this work well, defining jobs carefully and getting members to sign up for specific tasks with limited hour requirements, members will find the time. If you as a farmer find yourself in the awkward situation of converting from no work requirement, appeal to the members for assistance, explaining that you cannot cope with all the tasks a CSA entails. Several CSAs have used the list of CSA jobs from this book (see p. 65) to successfully activate their members. The GVOCSA sharers who come out to do farm work usually discover that they enjoy it a lot. While they cheerfully pick and pack our vegetables, I feel like Tom Sawyer whitewashing his fence. (See chapters 6 and 8 for more details on involving members.)

FAMILY WORK

Many CSAs are on very small farms, from one-half acre to ten acres in size, and with as few as half a dozen shares. The smallest CSAs are run by gardeners or homesteaders who are sharing their bounty with friends and neighbors. Calculations of hours and efficiency do not apply in the same way as in a commercial operation. When I moved to the farm in Gill, Massachusetts, one of my goals was to keep out of the supermarket. For eight years I grew, canned, froze, and bartered for my food supply. One year I even grew wheat on a raised bed. We scythed it down by hand, fed it through an old combine to separate out the grains, dropped the grain off the barn roof on a windy day to get rid of the chaff, and collected enough grain for five loaves of the best-tasting bread I ever baked. Was that efficient? Are you kidding? Did that matter? Of course not. It was definitely worth the effort to me and my son and his whole fifth grade class, who got to taste the bread with honey from our beehives.

Living on a farm blurs the line between life and work—it is a lifestyle choice that usually means less cash, fewer consumer amenities, and more physical labor. People should enter farming knowing this and should not be surprised and disappointed when they find they cannot earn as much as in most other professions or take paid vacations to exotic places. As in other small businesses, farmers have to pay for their own health insurance and plan for their own retirement. All the more reason, then, to learn how to work smart, and to use your own labor (and that of anyone willing to get involved with you) as thoughtfully and efficiently as you can. (Contact CSA Works for many suggestions on how to get more efficient.)

Farmers have been putting the family to work ever since Adam and Eve. The very idea of unpaid family labor makes agricultural economists squirm. Rarely calculated or remunerated in dollars per hour, that family labor is basic to the whole economy of an integrated farm, where one of the main "products" is the people. Sure, some people who grew up on farms look back in horror at their childhood slavery and swear, "Never again." They barricade themselves in a life of office work or whatever else is as far as possible from picking beans on the farm. But many more farm-raised people look back with nostalgia for the hours spent close to parents, remembering the many sensual pleasures of sounds and smells that folded in with the work. Doing chores and being expected to take responsibility is good for children. When they are really needed, they rise to the task. Old-fashioned as it may sound, having to work builds character. How else are we going to raise fewer couch potatoes and more active citizens? Learning that physical labor is good for the body *and* for the soul, and seeing parents involved in meaningful activity never hurt anyone.

Unfortunately, the economic pressures on small farms have all too often forced farmers to overload themselves and their families. Many love relationships have turned sour under the strain of picking too many vegetables for market. Should we then conclude that the work itself is the problem? Or rather that we need to change the economics and learn how to organize the work better? In a conversation about the push to save people from the drudgery of small farms, Wendell Berry exclaimed:

> Farming is a hard life. That's what these rural sociologists were talking about in the start. It's a hard life, therefore nobody ought to live it. What a remarkable conclusion. There are several steps that are left out. What causes the difficulty? Does freedom come out of it? Does family pride come with it, family coherence? Does some kind of idea of community come with it? Does some kind of idea of stewardship, of essential, irreplaceable, indispensable stewardship, does that come with it? Do ideas of affection or love or loyalty or fidelity come with it? The basic question is, how hard would you be willing to work in order to be free? (From an interview conducted by Elizabeth Barham in 1996 as part of her work on her dissertation.)

cosmos and pigweed

sometimes the work
seems no longer to be blessed
by companion hum of honey
bee or caress of the
sun,
hawk whistling high circles
on the breeze
but becomes more
mosquito whine and
nettle sting, sweat
in the eyes.
and where the mind once envisioned
flowers like stars
amidst crystal abstractions of
fiber and light,
and the eyes rejoiced
at broccoli and cosmos,
kale, calendula, lettuce,
and fennel—
it all becomes pigweed,
potato beetles, and the prickly mystery
weed that tears open the palm
no matter how calculated the approach.

what to do then?
find a tool.
keep your sights close.
rest, quench thirst, satisfy hunger.
take a walk.
admire your progress.
sing. breathe. stretch.
make conversation with the pigweed,
the potato beetle, the prickly mystery.
labor in love, not in vengeance, not in heat
of contest. not to win. just to do.
remember the flowers.

—SHERRIE MICKEL, 1992

There are volumes to be written about men and women working together on farms. Figuring out a fair division of work and responsibilities is extremely challenging. I realize this is a vast oversimplification, but years of observation have convinced me that most men and most women really do not think about the same things or in the same way. Ideally, this should be a source of strength for a working partnership where the masculine and feminine approaches complement one another. In reality, too often these differences are a source of friction. I have talked to farmers who run successful farms in partnership with close relatives. When I asked them what was most important for working together, their answers were similar: the ability to forgive and to place the family ahead of the farm.

At Rose Valley, David and I divided work in a pretty traditional way. David took charge of machines, field preparation, and short- and long-term maintenance. He is a skilled carpenter and has done enough tinkering and repairs on machines to have become good at it. My father could not even change a fuse, so I grew up innocent of most mechanical skills. What I know, I have learned painfully as an adult. On the farm, I planned crop production, ordered seed, kept the books, ran the greenhouse, arranged for marketing (including participation in the GVOCSA core group), and oversaw picking, packing, and delivering. When needed, I drove the tractor for cultivation and rock-picking. Both of us hoed and weeded and supervised apprentices. Twice a week, I worked with CSA members. David joined in on Sundays when we had a bigger crew than I could manage alone. Having clear lines of responsibility reduced conflict. In the spring, when David thought the ground was dry enough to disc, he followed his instincts and did not have to consult with me. I set the standards for quality in picking. David made no bones about my being fussier than he, but he accepted my standards and urged them upon our apprentices. In many ways our skills *were* complementary. What went wrong? I guess what so often goes wrong when the personal and the professional become inextricably entangled: a chance comment about the height of a shelf reverberates as a symbol of deeper levels of unhappiness and discontent.

According to Richard de Wilde of Harmony Valley Farm, near Viroqua, Wisconsin, he and his partner Linda Halley "are interchangeable." Richard does more than half of the actual farming, the maintenance, repairs, field work, and planting with their Stanhay planter. Linda helps him with crop planning, plants with their Planet Jr., does some of the tractor driving, and supervises the harvest crew. As the computer ace on the farm, Linda puts together the weekly CSA newsletter. The two share bookkeeping and take turns driving the delivery truck to Madison and serving as advisors to the University of Wisconsin in Madison.

It is interesting to compare the experience of Carol Eichelberger and Jean Mills, women partners. According to Carol, they share most work, but they divide responsibility according to what each cares more about, or dislikes the least. Carol is responsible for crop planning, pest control, the greenhouse, and irrigation. Jean takes responsibility for the organizational work, weed control, and harvesting, and writes the newsletter every other week. Together they do composting, tractor maintenance and repair, soil preparation, and green manuring. Carol says they have average amounts of mechanical know-how, but is pleased that their relatively new tractor runs well.

At Tierra Farm in New Hope, New York, Gunther Fishgold and Bruce Schader are two men partners. Their division of work responsibilities resembles what David and I did at Rose Valley. Bruce, who has been farming all his life, does the machine work and is responsible for farm operations. Gunther, with a background in environmental organizing and lobbying, takes charge of marketing, recruiting, and communicating with CSA members, as well as the crop planning, and works along with the apprentices on the list of jobs drawn up each week by Bruce.

A few years ago, Jean-Paul Courtens sketched out for Robyn Van En how many people it takes to run Roxbury Farm, with about twenty-two acres in vegetables and berries supplying 450 shares and limited non-CSA sales:

1 full-time farmer	2,500–2,800 hours/year
3 apprentices	1,400 hours/year/each
2 part-time coworkers	700 hours/year/each
1 bookkeeper	100 hours/year
3 site coordinators	40 hours/year/each

Total labor need—8,820 hours, including trucking and newsletter, but excluding volunteer help.

HIRING HELP

If you decide to hire workers, you enter a maze of state and federal regulations. As Doug Bowne put it at a workshop on this topic at the 1997 Northeast CSA Conference, you are immediately answerable to six or seven agencies that do not have clear answers for CSA farms. "We are pioneers in a litigious system," Bowne warned. For those who have never had employees before, the first stop should be the Internal Revenue Service, Circular A, the Agricultural Employer's Tax Guide. This document will lead you through the steps you must take. You can call 1-800-TAX-FORM and, on a good day, they will send you what you need. In many states, the Cooperative Extension or the nearest office of the state labor department will be able to provide you with information. The simplest course for getting started might be to consult with a neighboring farmer who handles the paperwork for his or her own employees. Some states have a tax guide for new businesses.

For all employees earning over $150 a year, the employer is responsible for social security and medicare. The combined tax rate is 15.3 percent of gross cash wages (this does not include payments in kind, such as farm produce or lodging). The employer deducts half from employee gross wages and pays the other half. To determine how much money the employer must withhold for the Federal income tax, new employees will need to fill out W-4 forms. The employer is then required to deposit all of the Social Security, Medicare, and Withheld Income Tax at an authorized bank. A special deposit coupon (Form 8109) must accompany these deposits. By January 31 of each year, employers must file Form 943, Employer's Annual Tax Return for Agricultural Employees, and send a W-2, Wage and Tax Statement, to each employee. Each state also has its own peculiar tax requirements. Calling the state tax number should furnish the appropriate regulations and forms. Employers must fill out an I-9 form for each employee to document citizenship, legal alien, or visa status, and keep it on file for at least three years. Federal law requires that all employers must also purchase Workers' Compensation insurance, which varies in price from state to state. Although the rates are set at the state level, the various kinds of farms have different risk pools. An uninsured employer is liable for the costs of medical treatment for a job-related injury or illness, and risks fines for failure to carry insurance. To fill out the many forms properly, you must have an Employer Identification Number (EIN), which functions like a social security number for tracking you through the system. To get one, fill out IRS Form SS4. The process can take a few months. Special laws regulate the hiring of youngsters under sixteen years of age who are not members of your immediate family. You must make sure they have working papers, and that you do not ask them to do jobs that are forbidden by law, such as driving equipment. The fines for violating these regulations are very heavy. Putting your own children to work may not look like a bad alternative after all!

Harmony Valley Farms
Viroqua, Wisconson

Finding really good, steady help is not easy. The pool of U.S. citizens willing to do farm work is shockingly small. The ideal employee is a local person whom you can train and keep for a long time. At Rose Valley, we were fortunate to find a friend, Greg Palmer, who lives nearby, and who began as an apprentice and graduated to junior partner status. Few CSAs are large enough to support more than one or two full-time workers. Even the farmers are not getting paid for year-round work. At Harmony Valley Farm, where almost four hundred CSA shares amount to half the farm sales, a man and a woman work year-round in addition to the two farm partners. The employees are local people who have been with the farm for many years. The summer crew of fourteen or fifteen is a diverse assortment of local teenagers, a few people in their thirties, some young Mexican guys who live in the area, and others who come from farther away. There have been Laotian Hmong, and in 1998 a man came from Ghana to learn organic farming. A few CSA members have worked for the summer, including one who is thinking of quitting his job as a college teacher and making a permanent move to the farm. The total payroll is $100,000 at a pay rate of $6 to $8 an hour. Smaller farms will be envious to learn that Harmony Valley also employs a full-time cook, an extern from the New England Culinary Institute, who, besides meals, supplies recipes for the CSA newsletter.

At Good Earth Farm in Weare, New Hampshire, with the help of one adult and three teenage workers, David Trumble supplies eighty families from three acres of vegetables and an acre of fruit. His experience in hiring workers has been very positive:

> We have had the privilege of employing a couple of dozen people over the years and maintain good friendships with all of them. They come back to visit us during college vacations, and have helped us babysit our child. We follow them to big events in their personal lives, or even just watch them in countless theatre productions at the local high school. We pay all of our workers a decent wage, deduct taxes, provide workman's com-

> pensation, take the time to teach them the skills they will need to be good farmers, and treat them like adults in the workforce. We are doing this with local labor, and we feel that this meets the definition of community better than importing interns from far away. We impact them, their friends, and their families. In addition, adults who stay with us for any length of time become like partners in that we seek out their advice and opinions in matters of agriculture.

Four partners share in owning and operating Full Belly Farm in Guinda, California, another larger farm where 600 shares make up only one third of the income. Four or five other people live and work on the farm and fifteen additional local people work year-round. Full Belly also hires a full-time person to coordinate and handle the bookkeeping for the 450 shares delivered to Berkeley. The partners take turns overseeing the daily harvest.

To help him farm the thirty acres he rents from the Stony Brook-Millstone Watershed Association in Pennington, New Jersey, Jim Kinsel hires three full-time seasonal workers. Since 1993, Jim has developed a close working relationship with one family from Zacatecas, Mexico. He pays the eldest brother, David, $12 an hour for a sixty-hour work week as his field manager. A younger brother, back for his second year, will start at $6 an hour in 1998, and manage harvesting. The youngest brother, who is "coming for the fun" for the first time in 1998, will start at $5.50 an hour. These starting salaries allow Jim to give raises over the season. Jim also pays state unemployment and disability fees of $3,000 a year, and provides housing in the farmhouse he rents from the Watershed Association. Jim reflects that "at one level, I'm tearing families apart—two of the brothers have wives—but I'm also bringing them together. The three brothers hadn't been together for a decade because the eldest was out of the house and working before the youngest reached maturity."

When I asked Bill Brammer, whose CSA has grown to eight hundred shares and is still expanding, whether his members do any work, he said, "Having all those people at the farm would drive me crazy." Then he told me that he employs a farm manager, two office workers, four foremen, and up to 110 field workers, most of them Mexican. In other words, he and I supervise about the same numbers of people. If I had to be boss to that many, it would drive me crazy! So you pays your money, and you takes your choice.

INTERNS

Taking on interns or trainees is an appealing way to get farm help for less cash (because of the legal meaning given to the term apprentice, intern is more appropriate for here). To do it fairly and well, however, means assuming a major responsibility to teach and share what you know. If you do not want to spend time explaining why you are doing what you are doing on your farm, and just want to do a quick training and assign tasks, you should not entertain the idea of interns. In the draft version of "Internships in Sustainable Farming: A Handbook for Farmers," Doug Jones sums up the difference between hired workers and interns:

> Workers usually do specialized work in one area of the farm; they often have prior experience; they receive an hourly wage and usually do not live with you. State and federal govenments have many regulations and officials assigned to protect workers from exploitation by employers. An essentially adversarial relationship is assumed to exist.
>
> With interns, on the other hand, you have a much greater obligation to instruct. They expect you to explain the "whys," not just the "hows." They deserve a diversified learning experience through a broad exposure to many different tasks, as well as through frequent discussion of the overall goals, methods, and systems of the farm. They are preparing themselves for a vocation, or at least

learning how to grow their own food. Interns usually live on the farm, expect to interact socially with you, and may wish to learn a variety of rural living skills. . . . In most arrangements, interns receive a cash stipend that is not directly related to the number of hours worked. Hopefully, they will share some of your ideals and aspirations, and a mutual rather than adversarial relationship will prevail. (p. 2)

[Note: This is actually a good description of how to achieve friendly and productive relations with employees as well.]

So, if you are willing and able to teach while you work, interns may be the answer to your labor shortage. In my experience, when you find the right people, interns not only lighten your workload, but also make the work itself more pleasant. Working with someone who is enthusiastic about learning what you are doing can transform an otherwise routine task. When you explain to your intern how a repetitive job fits in the context of the farm's systems, you keep alive for yourself the interconnections that are so essential to a sustainable farm. I still keep in touch with most of my former interns. Several have gone on to farm on their own, and four run CSAs, so our paths continue to cross in mutually beneficial ways.

Taking on interns has a broader importance as well. If we want CSA to continue to grow, experienced farmers must learn how to pass on their knowledge. "We need thousands, tens of thousands of new farmers!" a quote that could have come from Robyn Van En, but was part of Paul Bernacky's presentation on apprenticeships at the 1997 Northeast CSA

COURTESY EARTHCRAFT FARM

Tom and Sharon Smith set out broccoli on a mechanical transplanter while Jim Rose drives the tractor at Earthcraft Farm in Indiana.

Conference. The practical skills of farming, and, in particular, managing a CSA farm, are not part of the curriculum of many schools. A few institutions are starting to catch up with the encouraging demand for this kind of learning. In the meantime, those of us who are doing it need to train an entire generation from scratch in local, sustainable food production and all the skills involved.

Finding really good, steady help is not easy. The pool of U.S. citizens willing to do farm work is shockingly small. The ideal employee is a local person whom you can train and keep for a long time.

Most of the people who are interested in learning to farm come from non-farm backgrounds. They have a double set of lessons ahead: learning how to do physical labor, and learning how to farm. Shane LaBrake, who runs a two-year internship program at Ecosystem Farm sponsored by the Accokeek Foundation in Maryland, divides these differently into three sets of challenges: physical, mental, and emotional. As Cass Peterson puts it in her letter to potential interns, "It takes time to learn some of the basic farming skills that most people nowadays think of as 'unskilled labor.' It isn't unskilled labor. Hoeing requires agility and practice. Harvesting requires judgment and speed. Marketing requires communications skills and experience."

Getting into physical shape is only the first necessary step. To use your body without hurting yourself and perform tasks with the least amount of energy takes practice and experience. Cooperating with other people on physical tasks adds another layer of complexity for which most people are not born with an instinctive awareness. When I bend to pick up a board, I can sense instantly if my companion understands what is required to move that board in the easiest way. If you grew up in the suburbs as I did, you might need a few pointers to master what farm kids take for granted as common sense.

Shane's goal is not just to train farmworkers, but to empower people to farm on their own. To do that, he believes, requires the acquisition of skills, but also of confidence. The first year of his training focuses on skills, physical and mental. He has designed the second year to provide the interns with management experience and more confidence in their ability to recognize and solve problems. Long hours, low pay, fatigue, droughts, floods and pestilence—the emotional challenges of farming come along with the territory, built in to every real-life farm. Probably the only way to know if you will be up to meeting them is to try it out. At least, through an internship, you get to do it on someone else's dime.

In recruiting interns, it is essential to give a realistic picture of what working on your farm will be like, and to dispel the many romantic notions abroad about the idyllic farm life. The Angelic Organics brochure inviting interns stresses the team discipline involved in farm work:

> A farm is a weaving. Everything that happens on it affects everything else on it. If time is lost because of a late start, or using an ineffective tool, or because a communication is misunderstood, the work still has to be done some time; it doesn't just go away. It will have to be done in the afternoon or in the evenings, or we will have to hire more help. Otherwise, the weeds will get away from us, or the harvest won't be completed before the rain, or transplanting will be delayed and the crop will be impaired, causing our CSA members to receive shabby or inadequate produce.

Cass Peterson does her best to scare away the unrealistic:

> Farmwork is hard. It involves long hours at times, as well as sore muscles, insect bites, sweat and dirt, and the duress of cold, heat, and rain. Not

> many jobs require real physical strength, but you will need a good measure of endurance. If you are expecting summer camp, complete with weenie roasts, hay rides, and romps in the ol' swimming hole, please consider doing something else with your summer. We strive to have a good time, and we do like to kick back at the end of the day with a cold beer. But our success depends on a certain obsessiveness about what we're doing and a serious adherence to demanding schedules.

To make an internship a success for both farmer and intern, both sides need to be clear on expectations and responsibilities. In his handbook for farmers, Doug Jones provides a useful set of guidelines for both parties. Many farms ask that candidates fill out application forms, including essays on why they think they want to farm, work references, and a date for a personal interview. Sam Smith, who has worked with four interns a year for over twenty years, stresses fostering a sense of teamwork, giving the interns specific areas of responsibility so they have the experience of making some decisions, taking the time to walk the farm together every week and share observations, drawing up a weekly workplan, and holding weekly study groups on important topics. He and Elizabeth have a sort of final test for their interns (and themselves!)—they go away for a week, leaving the interns to run the farm on their own. In my experience, interns who have not farmed before need six to eight weeks to develop a clear idea of where they want to focus. At this point well into the season is a good time to ask them to draw up a learning contract, outlining what skills they especially want to acquire. At the end of the season, you can do a mutual evaluation of how well both sides fulfilled the contract, the intern through learning and the farmer through teaching.

"Interns expect you to explain the 'whys,' not just the 'hows.' They deserve a diversified learning experience through a broad exposure to many different tasks, as well as through frequent discussion of the overall goals, methods, and systems of the farm."

Many of the farmers who hire interns worry about what to pay. The law requires minimum wage and all the other legal requirements for any workers, including social security, medicare, and workers' compensation insurance. That does not compensate the farmer in any way for the value of the education provided. Ironically, you are less likely to get in trouble if you *charge* the interns than if you *pay* them. Doug Jones's research seems to have unearthed a solution. An agent of the New York Department of Labor told Doug that if an intern and the farmer sign a work agreement that lays out what the intern will receive, the department would consider that to be a waiver of minimum wage by the intern. Doug was also referred to a clause in the New York minimum wage law that allows you to pay less than minimum wage to "a trainee enrolled in an organized vocational education training program in agriculture under a recognized educational, nonprofit or governmental agency or authority . . . provided such program is approved by the Commissioner." Other states may have similar openings for organic farming groups to create training programs.

There are many internship listings and programs around the country through which a farm can advertise for candidates, and potential interns can seek placements. Appropriate Technology Transfer for Rural Areas (ATTRA) has assembled the most comprehensive list of both individual farms and organizational programs. If you call their 800 number (1-800-346-9190), they will send you a copy in short order. Their listing includes CSA farms, though that is not their focus. Only a very few programs are dedicated

CLEMENS KALISCHER

Interns harvesting chard at Caretaker Farm.

to training CSA farmers in particular, though many CSA farms accept interns. Shane LaBrake runs the two-year internship referred to above; Happy Heart Farm in Colorado offers a three-year program; the Michael Fields Agricultural Institute Internship Program gives a seven-and-a-half-month training, combining classroom study with hands-on projects. A group of the students at Michael Fields have started Stella Gardens, a CSA that they run under the guidance of their mentor, Janet Gamble. (See the CSA Resources section for more details.)

A model program, which I hope will be replicated in other parts of the country, is the Collaborative Regional Alliance for Farmer Training (CRAFT). As the name suggests, CRAFT is the creation of a group of farmers in eastern New York and western Massachusetts, many of whom run CSA farms. According to Katy Smith, the program arose from the two hours a week (from 6 to 8 A.M. on Saturday mornings) that Jean-Paul Courtens committed to sharing his knowledge with her while she was an intern at Roxbury. In Katy's words, "I became increasingly aware what a big burden this was for Jean-Paul, every week to prepare two hours of information just for me. . . . I feel very fortunate! After two years, I was able to go out and start my own business, which has been successful. But it was draining for him." To lighten the load, they began visiting other farms in the area. In the winter of 1995, the farms involved met and agreed to exchange farm visits to provide all of their interns with a broader exposure to the variety of farming techniques and designs even among organic and Biodynamic farms. Sam Smith, one of the founding farmers, writes that a second, but equally compelling purpose, has been "to enable the apprentices to experience the importance and power of forming close associations and a supportive community among peers" (*CSA Farm Network*, vol. II, p. 45).

Each of the thirteen farms participating in CRAFT hosts a tour, and the farmer speaks on a favorite topic. Sam reports: "For the 1997 season, the topics are: tractor maintenance and farm safety, integrating livestock and crop production, hands-on experience at a dairy farm, cover crops, tillage, and water management, getting started, soil structure, financial management, seed production and saving, orchard management, greenhouse operation, and soil fertility and testing." The sessions last a half day every other week: they tried full-day and two-day sessions, but those turned out to be too exhausting. The program opens and closes with general gatherings for introductions and evaluation. From all reports, CRAFT has been a great success in enriching the learning experience of the interns while strengthening the sense of community for both them and the farmers. One of Sam's interns observed, "It has been reassuring to discover that my farmers, Sam and Elizabeth, are part of a much larger circle of farmers sowing, cultivating and sharing their beliefs" (*CSA Farm Network*, vol. II, p. 46).

WE CAN EXPERIENCE FARM LABOR as an ecstasy of oneness with the Earth. Josh Tenenbaum, a member of GVOCSA wrote: "When we dig our hands into the brown, damp soil, does not the entire Earth tremble? When we plant seeds and eat green, living food, does not all the Sun's light intermingle with our own? We see that planting and weeding, building a community around sustainable agriculture is the most fundamental peace work, no matter what our political beliefs or ideas" ("Global Effects of CSA Participation," 1991). Or we can experience this labor as spirit-crushing, unrelenting drudgery. The challenge is ours to make of it what we will.

SHARERS ON THE FARM 8

The only possible alternative to being either the oppressed or the oppressor is voluntary cooperation.

—Enrico Malatesta

Time-study experts will tell you that the most efficient harvesting system gets the most vegetables cut, washed, and packed in the shortest amount of time by the smallest number of people. If you aspire to this definition of efficiency, you will have a tight, experienced crew do all of your harvesting, as they do at the Food Bank Farm in Massachusetts and at Moore Ranch in California. However, you might want to consider other values, such as community, education, and participation. If you rank these higher than efficiency, and you are sociable by nature, you may decide to involve your sharers in the harvest.

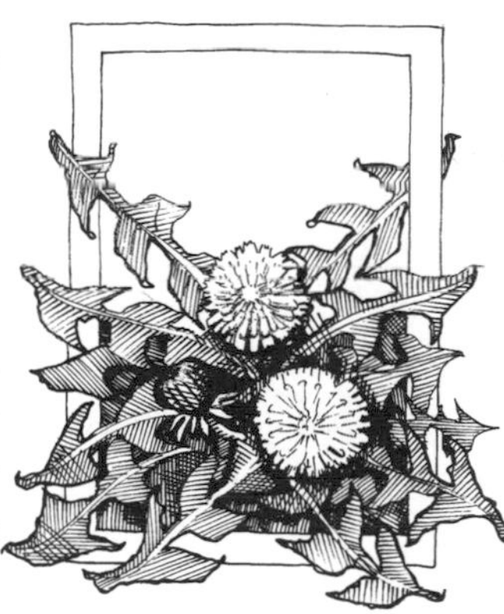

As Robyn Van En put it, "the weekly harvest days are a great time to engage CSA members in hands-on activity. The members should always be accompanied by the regular farm crew. It is a wonderful opportunity for them to bond with the farm or garden, as there is little to compare to mornings in a field, the crops heavy with dew, and the birds singing. Members who feel welcome and take the time from their regular work to help with the harvest often become regular volunteers. Deep, philosophical conversations can arise while everyone is vigorously picking string beans. The community spirit is palpable. Many hands make light work and create an unavoidable feeling of accomplishment.

"If your project does plan to have volunteer or mandatory help with the farm work, make member participation as easy as possible. At an orientation meeting or in the newsletter, list all the jobs that need to be done. Indicate the days of the week and hours of harvesting, critical times in the season for weeding, setting out transplants, and the like. Let members know which jobs are 'down an' dirty' and which are stand-up with an apron on." At GVOCSA, we sign members up for "Special Vegetable Action Teams" (SVAT) for jobs like planting onions, putting up the pea fence, and harvesting winter squash or garlic.

Every Community Supported Agriculture project has to decide whether to involve members. Some

CSAs do not ask members to work at all. More frequently, CSAs offer the choice of a reduced fee for working a certain number of hours per season. Only 22 percent of the CSAs Tim Laird surveyed *require* that their members pitch in; many more solicit volunteer help. The GVOCSA requires work as part of the contribution for every share, as do Blackberry Hills Farm in Wisconsin (one work day per member), Casey Farm in Rhode Island (eight hours per share), Holcomb Farm in Connecticut (four hours per individual share and fifty hours per organizational share), Many Hands Organic Farm in Massachusetts (four hours per share), and Valley Creek Community Farm in Minnesota (one harvest or delivery per share). The Genesee Community Farm in Wisconsin functions more like a cooperative, with each member working three hours every week.

According to Laird's study, work hour requirements range from three to eighty hours, with an average of twenty. The completion rate for required work was 70 percent, with four of the fourteen farms claiming 100 percent and the others ranging from 5 to 90 percent. Some of the farms do not require work, but instead offer some crops on a pick-your-own basis. Where members come to the farm for pickup, inviting them to pick for themselves is a popular approach for high-labor crops, such as peas, beans, berries, and cherry tomatoes (Laird, "Community Supported Agriculture," pp. 54–55).

Casey Farm
Saunderstown, Rhode Island

AT QUAIL HILL FARM in Southampton, New York, members pick most of their own vegetables. A video of Quail Hill on a picking day shows a remarkable tableau of dozens of people harvesting asparagus and spinach. When interviewed in front of the camera, the members express their delight at the opportunity to pick for themselves. To prepare members for this harvest, the farm crew holds three field orientation days in the spring. They teach members where to find the crops, how to pick or cut, and how to walk around a bed instead of through it. Veteran members are usually willing to help newcomers, but the crew is always available during harvest hours to answer questions. Maps, clear markers, and signs help prevent chaos.

ONE CSA FARM has a labor credit system offering members $6 per hour toward the price of their share. West Haven Farm at the Eco-Village in Ithaca, New York, gives free shares to members who help harvest for two months of the season. Live Power Community Farm in Covelo, California, ships its shares to San Francisco, over four hours away. Members rarely come to the farm to help work, but have full responsibility for distribution in the city. Ruckytucks Farm in Stillwater, New York, gives members the choice of working, but sends those who don't show up a bill for the hours missed. New Town Farm in Waxhaw, North Carolina, asks working members to pay the full $375 for a share and then gives them a $50 rebate when their work is completed.

AT SILVER CREEK FARM, the Bartletts entrust their forty-five working core members with carrying out both the harvest and distribution. For $100 off their share fee, the working core puts in eight half days for the twenty-two-week season. There are two coordinators for each work session. On Wednesdays they harvest, on Thursdays oversee distribution, and on Sundays do weeding, planting, or whatever farm work needs to be done, including preparation for other markets. An intern works with the members, while Ted and Molly are free to do other tasks. Work begins at 8:30 A.M., the members

ELIZABETH HENDERSON

Genesee Valley Organic CSA members cooperate in dismantling the pea trellising at Rose Valley Farm.

are welcome to bring their children, and the morning ends with a picnic lunch and a swim in the farm pond. One woman has become such a regular that, for the 1998 season, the farm will pay her with a free share and a ten hours per week salary to coordinate the other core workers.

During our first year of operation, our little core committee of David and me (the farmers), and two consumers, decided to require farm work as an experiment. We were not sure people would like it, but we wanted to try. We reasoned that people who could afford to buy their way out of work were already being served by the local food co-op and health-food stores. To keep CSA prices lower, members would help administer the project and provide labor, especially for picking and packing. For people who could neither afford to shop in stores, nor do the work because of illness or physical disability, we organized a small buying club. The coordinator received produce in exchange for taking orders, collecting payments, and using her home as pickup point. After nine CSA seasons, with a membership that has grown to 160 shares, we can say confidently that requiring farm work has been a great success, enabling us to keep share price low—$9 to $15 a week on a sliding scale. Members consider the farm work a benefit. Their end-of-the-season evaluations are unanimously positive about only two things: the quality of the food and the farm work.

Our CSA distributes three times a week. At the beginning of the season members sign up for a particular day for farm work, distribution work, and weekly pickup. Two mornings a week, groups of sharers drive to the farm. We figure one helper for every ten shares is adequate for the amount of work required and for vehicle space to transport the food from the farm to the pickup center in a church in Rochester. Each share entails three 4-hour sessions at the farm, and two 3-hour sessions at the distribution center.

We prepare sharers for the farm work with detailed suggestions on what to wear and what to bring, depending on the weather. Sharers arrive dressed in boots and hats, carrying water bottles, sunscreen, and bug repellent. Some show impressive foresight, coming with work gloves, gloves for warmth, and a third pair of rubber gloves for washing vegetables; rain jackets, winter jackets, and a change of clothes if they get wet. Some of the women tell delighted stories about their reception at a coffee shop where they stop on the way home covered with mud. My heart sometimes sinks when I wake up to a CSA workday of particularly trying weather. One snow-covered October morning, I was starting to mumble an apology for the snow to the first comers, when a second carload arrived. Kim Christopoulos leaped out exclaiming, "Yeeha! It's too slippery to ride a motorbike. Let's get picking!" No apologies were needed after that.

Most members consider the farm work a benefit. Their end-of-the-season evaluations are unanimously positive about only two things: the quality of the food and the farm work.

After signing up, each sharer receives a copy of the work schedule for the entire season. Anyone who wants to change work dates is responsible for trading with someone else and reporting to the schedule coordinator for that day of the week. A few people are usually willing to be on call at short notice for emergencies. Frequently, extra people come or members bring friends and relatives. The attendance record has been phenomenally good. Only once in nine years has no one shown up on a workday. On that occasion, one CSAer suffered a detached retina the night before; the other two lost their brakes on the way to the farm.

We plan carefully for each farm workday, aiming to keep everybody busy at a relaxed pace. Mornings when only three or four sharers come, one of us works with and supervises them. When more people come, we divide into two crews. Usually, we spend two to three hours picking, washing, and packing the food for that day's shares. With few exceptions, totally unskilled people can learn how to make bunches of greens, dig carrots, and pick beans, peas, corn, peppers, tomatoes, or berries. The hardest part is getting the count right. We count out rubber bands in advance so we don't have to try to count as we pick. We (the farmers) usually harvest vegetables that require experienced judgment—asparagus, lettuce, broccoli—the day before or in the morning before the CSAers arrive. After everything is picked and stored in the cooler, we spend the last hour or so weeding or hoeing. In the fall, sharers clean and sort onions, garlic, potatoes, and squash. We've agreed with our insurance company that we will not use chemical pesticides on the farm, and that only the farmers will use machines or ladders.

New CSAers, who have never been on a farm before, or even gardened, are often nervous that they will make a mistake or step in the wrong place. We try to reassure them by explaining as we go along where things are and why we have planted the way we have. Most of the tasks are so simple and clear that people quickly realize they will do all right. After a year or two, some of our sharers have become old hands who need only the slightest direction and can be trusted to do tasks carefully. Obviously, they don't get the work done as quickly as a skilled crew, but our sharers have been remarkably conscientious and willing. I can think of only one man who was an exception. I noticed him doing a very sloppy job, mixing as many weeds as Swiss chard leaves in his bunches. I pointed out to him that he was going to eat that food. A light seemed to go on in his brain. He replied, "Oh, I see what you mean. You and

David are professionals, so we need to take pride in this work."

VALERIE VERVOORT, a member of Zephyr Farm CSA in Stoughton, Wisconsin, appreciates the socializing that can occur while doing jobs that can seem like boring stoop labor when done alone: "The bean field is where you really get to know people. There are so many beans on one bush, you're sitting on a bucket picking them forever. You get into politics and issues, everything from whether you like the governor to the role of men and women in society. And when the group is small, it can get pretty confessional about relationships" (*Country Journal* [May/June 1998]: 26).

AT HAPPY HEART FARM in Colorado, about 20 percent of the members work three hours a week in exchange for a 50 percent reduction in the share price. Dennis Stenson describes sharer help: "[when members work with you,] there is just a rhythm to the dance, rather than a frenzy, and the working members share that dance with us and our interns. Supervision and planning play key roles in maximizing the blessing of those helping hands; they are working with us rather than for us. Time must be spent with each new task and with each new worker to show how, where, and when, rather than to tell. Constant follow-up is also essential, because anything taken for granted will be a mistake" (*Seasonal News* 1, no.1 [1994]: 3).

CLEMENS KALISCHER

Sharer volunteers picking beans at Indian Line Farm, Great Barrington, Massachusetts.

The 1992 season at GVOCSA put to the test our members' agreement to share with us the risks of farming. Along with the rest of the Northeast, we suffered from a cold, wet growing season. To make things worse, our county was hit with disastrous rains, which began with five and a half inches in one afternoon, followed by twenty-five inches in thirty five days. We averaged three quarters of an inch of rain a day, the kind of aberrant weather that is predicted to accompany global warming. Half of our crops drowned outright while the rest eked out a third to a half the usual rate of production. In mud up to our boot tops, at the end of July, we asked the core committee if they would prefer us to give back the money and close down for the rest of the season.

They answered unanimously that they didn't care if we gave them only one item a week; they were determined to stick with us. The sharers were more worried about us than about how much food they got. They called, wrote encouraging letters, and even offered to donate money. They showed up on time for their work and slogged with us through the mud. After a morning with us, Sean Cosgrove wrote:

> Driving back from the farm today, we all commented on how much we enjoyed our visit—even in spite of the rain. I wanted to thank you again for your hospitality, for sharing your knowledge about growing things and for the work you are doing.
>
> I imagine that things may be looking a little bleak right now, but the CSA project is so important. I've been despairing a lot lately about our food system and the choices we are given—so it is reassuring and heartening to me to see that there

ELIZABETH HENDERSON

Genesee Valley Organic CSA sharers sorting and tasting corn at Rose Valley Farm.

> are alternatives . . . albeit struggling ones. Besides the delicious lunch, the fine company, and the pleasures of working a little bit on the farm, you gave me some hope today. So—I'm sending some hope back your way along with many thanks. I look forward to returning to the farm again—hopefully under brighter circumstances.

Our sharers kept us going. The 150 people who attended our end-of-the-season dinner gave us something not many farmers get to experience from the people they feed—a standing ovation.

Some of the farmers surveyed by Tim Laird expressed a fear that people would not join a CSA if required to work. Other farmers told of their disappointment at the low level of participation in voluntary farm work or of the nagging necessary to get members to fulfill their commitments. Switching from no work to a work requirement could be quite difficult. However, when members participate, their sense of commitment to the farm and appreciation for the farmers' work increase dramatically. As Dennis Stenson puts it: "We've found over the years that it is the working members who really get all the deeper economic and social concepts that CSA is about. Their families become the long-term, committed, enthusiastic, word-of-mouth sales reps we need to keep up with the annual fluctuation of those who come and go."

After her first work experience, Pat Mannix, one of the founding members of the GVOCSA, wrote this deeply felt appreciation:

> For four hours we worked. We picked broccoli, corn, beans, greens, and herbs. We washed and packed them for transport. We hand-weeded and hoed, making a way through the clumps of earth for tiny seedlings. The sun beat down, the mosquitoes bit, the sweat ran. My skin itched, my shoulders ached, my cheeks burned, my dry throat scratched. It was wonderful!
>
> During the week as I prepared meals, I noticed a different feeling, a change in perspective. I found myself preparing the vegetables in a loving, respectful manner. I planned with a passion so nothing would go to waste. I began to compost. When, for the first time, I ate what I had harvested, it was both an awakening and yet, a deepening of the mystery. I clearly understood how this food was becoming a very part of me; its life had been sacrificed that mine could continue and this was all right. It was as it was meant to be and this made it sacred. The old liturgical declaration—fruit of the vine, work of human hands—took on new meaning. I understood that the Earth was alive and that it gave and sustained other life. I knew that the vegetables and myself were both children of it, joined in a wonderful kinship. Food would never be the same for me again.

And for the farmers, the experience of community and support is worth the effort to involve members in the farm. After Julie Rawson's first season with seven working shareholders doing eighty hours each, replacing her four dispersing children, she wrote:

> I get a sense of confidence from so many people having ownership in the day-to-day operation. Interestingly enough, their presence made the experiences for the paying shareholders, whose work commitment was only four hours, more meaningful. It was with great peace of mind that we left the farm for a week in August and left the working shareholders in charge of the entire management. (*CSA Farm Network,* vol. II, p. 74)

Andrea and David Craxton at Roots and Fruits in Dalton, New Hampshire, wrote this appreciation of their members' work:

> We had a lot of return customers, which translated to lots of experience and expertise in field work. Increasingly, our operation is becoming a well-oiled machine. The old-timers quickly instructed and supported the newcomers as they

> adopted the procedures that we follow. We soon became a big, happy, growing tribe of seventeen families. Our Friday morning harvests were, for many, the highlight of the week. We can't say enough about the enthusiasm and good will as everyone pitched in and made our CSA a delicious success. (*CSA Farm Network*, vol. II, p 83)

For me at Rose Valley, the intimacy that came from sharing our farming transcended our business connection. Working with our customers is also enriching socially. I enjoy the opportunity the CSA provides to get to know the people who eat our food. While our hands perform routine tasks like picking and weeding, we are free to have long and intense conversations. Sharers get to know one another as well: surprising exchanges occur between an engineer and a welfare mom or a nurse-nutritionist and a newspaper reporter working together. Seeing the children at regular intervals year after year, we have the pleasure of watching them grow and develop. We don't just sell our sharers food. We were adopted by this growing community of friends who considered Rose Valley "their farm" and us "their farmers."

We reasoned that people who could afford to buy their way out of the work requirement were already being served by the local food co-op and health food stores. To keep CSA prices lower, members would help administer the project and provide labor, especially for picking and packing.

CHILDREN ON THE FARM

Children are welcome at most CSA farms. Willow Pond Farm has a special garden by and for the children. Marty Grady, an appreciative parent wrote:

> When we go to the farm, we take a picnic so we don't have to rush out and rush back. The first place we go when we get there is the children's garden. Each week my children monitor the progress of the garden and their favorite pumpkins. They're really involved in the whole growing cycle. Jill is very open—she really trusts the kids and their exploration. They're free to roam the fields . . . and to use the bathrooms! The kids just soak it up out there. It's great!

At Rose Valley, children were free either to participate in the farm work or to play in clearly designated areas: the sandbox, the swings, or the packing shed on rainy days. We had clear rules against touching equipment or wandering into sheds and buildings. Though we had as many as nineteen children at a time (for a birthday party), no one ever violated these rules. We encouraged families with small children to sign up for designated workdays for which we scheduled an extra adult to do childcare. Some families preferred to come an extra time so that the parents could take turns between childcare and farm work.

To start off a work morning with children, as part of our introduction circle, one of us was always sure to get the children's attention and explain our rules (see pp. 103–104 in chapter 10). Shy on their first visit to the farm, by their second trip, children have made it their own turf. Most children divide their time between work and play. Children as young as four or five have helped for the entire four hours with a break or two for snacks. Few children like weeding, but many take part in washing and sorting jobs, and pick beans, peas, corn, and berries with enthusiasm. One morning, when, at the last minute, a father could not make his work assignment, his wife persuaded their two sons, four and six, that they could replace him. The two boys pitched in willingly—they counted rubber bands, washed lettuces, and began picking beans. About halfway down the row, they turned to their mother and said, "You know we've done more work than Daddy would have done!"

Another morning, the presence of a large group of children inspired us to dig potatoes. If you have ever run a crew of teenage potato gatherers, you know getting them to work can be harder than picking the potatoes yourself. The CSA children, aged two to six, had never seen a potato harvest. As the digger brought the red potatoes up from the ground and dropped them in a row, the children shrieked with delight. From a distance one might have thought we were running one of the wilder rides at a county fair. The children ran after the digger, snatching those potatoes up as fast as candy out of a piñata.

Caretaker Farm includes clear guidelines for parents in their CSA Handbook:

The farm is a wonderful place for children to explore and learn about plants, animals, and where and how their food is grown. It is also a place to learn respect for the land that feeds and delights us. While we want the children to feel at home on the farm, we also want them to be safe.

Parents are expected to know where their children are at all times.

We have created some guidelines to be followed:

1. *The Ponds.* Children must be accompanied by an adult swimmer when they visit the ponds.
2. *The Electric Fence.* The fence is on at all times to contain livestock and to protect them and the gardens from coyote and deer. All wires are "HOT," including the temporary plastic fence. It is not dangerous to touch, but it isn't pleasant either. *We are most concerned about babies and toddlers who are too young to know.* Please educate your children to respect it.
3. *The Animals.* For the protection of the children and of the animals, please do not enter the animal pens, the chicken houses, or the pastures unless accompanied by a farm staff person.
4. *The Farm Buildings.* Barns, workshops, tractor sheds, and greenhouses are off-limits to children unless accompanied by a farm staff person or an adult.
5. *The Children's Garden.* This is a special place for children when they are visiting the farm. Families who wish to help may adopt a bed for the summer.

Weather conditions that discourage grownups may have the opposite effect on children. The truth is, if we will admit it, we all like to play in the mud. Children were especially appreciative of the opportunities afforded by 1992—the Year of the Big Waters. Nathan, son of Pam Hunter, the minister of the Lutheran Church which hosts our pickup site, was two that summer, toddling along on his hind legs, but just barely. Pam was apprehensive as he walked from the swing to join us at the beds where we were picking. No more than three feet up the row, Nathan did a perfect pratfall face-first into the mud. He struggled back on his feet, pushed back his rain hood and laughed. The worst had happened. Pam relaxed. At the end of another rainy morning, three older children lingered near the packing shed, reluctant to get in the family car. They crowded around me, faces streaked with mud, asking excitedly, "Do you get this dirty *every* day?"

The families in our CSA who are homeschooling consider participation in the farm work a significant addition to their children's education. Ann Bayer wants her four children to understand where their food comes from. Having grown up in the city, she does not want her children to believe, as she did, that food comes from the back room at the grocery store.

Members of other CSAs around the country have expressed how important the farm connection, the

Susanna Hamer at two and at eight harvesting potatoes at Rose Valley Farm.

pesticide-free food, and the education about how food is grown is to their children. Many report that their children eat more of the vegetables after they've been to the farm and helped pick them. In their sociological study, "Factors Influencing the Decision to Join a CSA Farm," Jane Kolodinsky and Leslie Pelch did telephone surveys of members and non-members of three Vermont CSAs, and came to the puzzling conclusion that having children makes people less likely to join. These "scientific data" contradict the real live CSAs I have known in which members with children make up the majority, and many parents say they have joined *because* of the children. A member of Harmony Valley wrote this comment, which I think typifies parent sentiment about CSAs: "The biggest benefit is that our kids could see where our veggies come from (at Strawberry Days), and I believe this encouraged them to believe that veggies are a wonderful gift and therefore would eat them."

MONEY MATTERS FOR CSAs 9

There can be nothing sacred in something that has a price.

—E. F. SCHUMACHER, *SMALL IS BEAUTIFUL*

BESIDES ALL THE WONDERFUL EARTHY and human qualities of CSAs, they must also function as viable small businesses. Throwing all the receipts in a shoebox might work for the first year, but long-term survival requires economic management systems for budgeting, pricing, bookkeeping, and contractual agreements with members. An essential aspect of CSA sustainability is providing a decent living for the people doing the farm work, and adequate cash flow to cover both the annual costs of production and the continuing reproduction of the farm, that is, investments for maintenance, improvements, transfer to the next generation, and retirement.

SHARE PRICING AND CSA BUDGETS

CSAs fall roughly into three groupings according to how closely price is linked to the amount of produce in a share. Even subscription-style CSAs weaken the market connection between the amount of produce received and the price paid, since they do not price items individually, nor necessarily maintain share size from week to week. The farmers charge a given weekly price for a share without divulging how they arrive at that price. Heavy weeks may compensate for light ones to reach a yearly average.

A second group of CSAs goes a little further towards breaking the price/produce connection. These farmers present a budget to the core group for discussion. To meet the budget, these CSAs divide the total figure by the number of shares. CSAs concerned about economic diversity permit sharers to pay on a sliding scale. Instead of charging by share, a few farms charge so much per adult member and feed the children for free. A very few community farms, like Temple-Wilton in New Hampshire, break the price connection completely. These farmers submit a budget to the members, who then pledge what they can afford to contribute to the expenses of maintaining the entire farm. If the first go-around does not cover the costs, members are urged to pledge more until the budget is either met or altered to adjust to their ability to pay. Regardless of how much he/she pays, each member takes as much food as he/she needs.

Trauger Groh, of Temple-Wilton Community Farm, is the eloquent spokesman for CSA as part of a new kind of "associative economy." By Trauger's logic, farmers are too busy farming to be able to bother with earning money, so other people must support them. He defines an associative economy as "an economy where the motive of our actions is the need of others, and that is what a farm is about—the motivation of the farmer is the need of the members for produce, and the motivation of the group is the need of the farmer." Or as Karen Kerney, the artist for this book, who runs a mini CSA in Jamesville, New York, puts it: "I believe when you grow food—you grow food. You do not make money. The money is made in the harvest and handling."

Share Prices

Regarding sign-ups for next year: We raised our price $15 per share. We are offering a $15 discount to anyone who signs up by December 10; as you can probably tell from how far ahead of time we are dealing with 1996, we want to sell our shares early. (You might also note: We are already preparing fields for 1996, so this forward planning does come out of the farming process.). Several of you suggested that we raise our price to cover the cost of production. I know that price would have prevented many shareholders from signing up again; it would have simply been too high. That's known as "pricing yourself out of business." The price we are asking might cause us to eventually farm our way out of business, but at least it will keep us going while we try to figure out a way to make Angelic Organics truly sustainable. We'll include a 1996 sign-up form with a future newsletter.

For 1996, we are creating different categories that shareholders can sign up for beyond the basic box price. We think this is an effective way to give you a choice about the long-term future of Angelic Organics. This method does not force a big price increase onto shareholders, but it does create a clear opportunity for people to pitch in to make this farm continue. The different levels of support are: seed $15; cotyledon $25; seedling $50; vine $100; pumpkin blossom $500; giant pumpkin $1,000. You'll get this information when you get your form later in the season, but I'm including it in now, so you can mull it over.

We understand that for many, renewing a share at the basic share price is a considerable financial stretch. If you want to help out Angelic Organics, but money is not your way to do it, another way you can pitch in is to encourage your friends to join Angelic Organics for next year—sometime this fall would be best. We hope it seems more like fun than work for you to get your friends to join Angelic Organics; then you get to compare recipes, carpool to our field days, trade vegetables, and keep your friends healthy! For us, it is very costly and time-consuming to persuade newcomers to join our CSA, when we have our hands (and hearts) full of farming. Most people seem stuck in the paradigm that anyone they buy anything from is just another business. If we were just another business, first of all, we wouldn't be in business, and secondly, we'd have a boring newsletter. If you think your friends would like what we do, please encourage them to join—this makes a huge difference to Angelic Organics. Note: we only want people to join who are aligned with the CSA model; it's impossible to please people who think of us as a store. We're a farm, through and through.

—John Peterson, Angelic Organics, Caledonia, Illinois. Used by permission.

CSAs also vary in how public they make the budget. The earliest North American CSAs, following the model of Indian Line Farm, recruited members by circulating a brochure that included the budget. In the Northeast, some CSAs continue this practice, but it is rare in other parts of the country. When Rose Valley Farm launched the GVOCSA in 1989, we did not share budget information with the general public or even with our core group. The thirty-one shares represented a small percentage of our farm sales, so we did not feel discussing our farm finances with the members was necessary. Of the CSA farms Tim Laird surveyed in 1993, 74 percent sold to multiple markets. In 1996, when we cut back on production and dropped most other markets, we printed our farm budget for the first time in the GVOCSA newsletter. The certification fee was the only item the core group discussed at any length. The $650 fee was disproportionately high because it was based on our previous year's sales. Deciding that continuing to certify was important, the core agreed to ask the members for additional contributions to help cover the fee.

By contrast, the farmers in the Temple-Wilton CSA analyze their expenses at length with their farm members. From the beginning, they have conceived of their project as a community farm in which all members share tasks, responsibilities, and financial contributions according to ability. Not many CSAs have taken this path. Like Rose Valley, most farmers initially view the CSA as an alternative marketing approach for the farm. Only gradually, as the GVOCSA has become a genuinely supportive community and a larger part of our marketing, have we begun sharing more about our farm as a business. Michael Docter of the Food Bank Farm takes the pragmatic attitude that members are more likely to be concerned about the value of the food they receive than about the finances of the farm. He contends that, to become widespread, CSAs must produce food efficiently and provide good value to their members. Many other CSA farmers would appreciate more membership involvement and support, but feel they must move in that direction cautiously so as not to scare members away.

Whether you make it public or not, the farm budget needs to include certain standard expenses. The county Extension office or the Farm Service Agency can supply you with budget examples. The Internal Revenue Service in Schedule F (Form 1040), "Profit and Loss from Farming," provides a handy guide. Setting up your bookkeeping with this form in mind will aid you in doing your taxes. Here is the IRS's list of legitimate farm expenses:

- car and truck expenses (form 4562)
- chemicals
- conservation expenses (attach form 8645)
- custom hire (machine work)
- depreciation and section 179 expense deduction not claimed elsewhere
- employee benefit programs other than on line 25
- feed purchased
- fertilizer and lime
- freight and trucking
- gasoline, fuel, and oil
- insurance (other than health)
- interest
  a. mortgage (paid to banks, etc.)
  b. other
- labor hired (less employment credit)
- pension and profit-sharing plans
- rent or lease
  a. vehicles, machinery, and equipment
  b. other (land, animals, etc.)
- repairs and maintenance
- seeds and plants purchased
- storage and warehousing
- supplies purchased
- taxes
- utilities
- veterinary, breeding, and medicine
- other expenses (specify)—Includes investments in land improvement, such as tiling of fields, gifts and contributions, legal and professional fees, advertising, dues and seminars, (including certification), pest control, and others.

Note that the IRS does not include a salary or health insurance for the farmer. In official business tax-think, the farmer receives the difference between gross revenues and expenses—the "profits."

FARMER EARNINGS

As Jill Agnew of Willow Pond Farm puts it, "Most of the food in this country is provided by people working for substandard wages. Paying people a living wage for work in agriculture is also critical to the CSA philosophy." Published budgets usually include the farmer's salary along with other labor costs. In a few cases, CSA members have been shocked at how little the farmer earned and have insisted on raising share fees to provide a living wage. Both Tim Laird and Daniel Lass discovered that many of the CSAs they surveyed did not include farmer salaries or capital expenses in their budgets. Among Laird's respondents, the average salary was $11,225, with a range from $0 to $30,000. Laird discerned a correlation between population base and salary amount:

> CSAs that draw on populations above 500,000 people average a salary of $16,625 (subtracting three farmers whose salary is zero), while farms with a population range between 100,000 and 500,000 people average $10,100 (with no growers stating a salary of zero). Farms with a population range between 25,000 and 100,000 average a yearly salary of $8,300 (subtracting four salaries of zero). Five of eight growers with a population base of less than 5,000 have a salary of zero. (Laird, "Community Supported Agriculture," p. 79)

Budgets in CSA brochures usually divide expenses into labor, capital, and operating costs. Labor includes salaries for the farmer or farmers, assistants or interns, additional hourly help, and insurance and tax costs. I have seen only a few budgets that build in a pension fund for the farmer. Operating costs may be divided into separate sections for farm, share distribution, and CSA administration. The published budget for Happy Heart Farm in Colorado for 1996 included a shortfall to be made up by fundraisers. Under capital expenses, CSAs include the purchase of tools, equipment, fencing, and construction costs for coolers, and other durable materials. Large expenses, such as tractor purchases and greenhouse construction, are usually amortized or factored in over several years (see examples, p. 93–94).

CAPITAL INVESTMENTS

To avoid taking it out of their own hides or borrowing from banks for major improvements or expansion, several CSAs have appealed to their members. As Signe Waller and Jim Rose of Earthcraft Farm wrote in a fundraising letter to the farm's most loyal members: "to be a sustainable operation, Earthcraft Farm needs a subsidy from a source other than ourselves." Along with the annual budget, they included "Items to be Purchased from Sustainability Fund," which was a list of soil amendments, livestock, and equipment that will contribute to building the long-term fertility of the farm. Happy Heart cheerfully invites people to sign up as "Friends of the Farm," holding out the "opportunity to provide the needed *capitalization* funding." Their brochure lists four levels for donations: Root—$25, Stem—$50, Leaf—$100, and Flower—over $500. Another Colorado farm, Blacksmith Ridge, has a similar scheme. In their 1996 brochure, Harmony Valley asked members to help the farm amass money for a down payment on land for a permanent site by prepaying for 1997 and beyond. Members came through with enough money, but the owner of the land the farm wants is not yet ready to sell. In 1997, Angelic Organics invited sharers to invest in a limited liability company to purchase 38 acres of additional land for the farm. (See chapter 5 for examples of member contributions to land purchases and conservation easements.)

TENTATIVE 1997 BUDGET AND INCOME PROJECTION

Anticipated Expenses:

Equipment repair		$2,000
Fuel		2,500
Greenhouse		3,000
General supplies		2,000
Labor & stipends		25,000
Farmer	$13,000	
2 Interns	6,000	
Driver	1,500	
Indianapolis organizer	1,300	
Hired labor	3,200	
Memberships		500
Miscellaneous		500
Office		1,800
Rentals		800
Soil amendments		700
Seeds		4,200
Farmers' salaries (2)		28,000
Tools & equipment		12,000
Corn planter (2-row)	500	
Potato digger	2,000	
Other	9,500	
Taxes		2,000
Utilities		800
Vehicle		1,000
Debt service (see loan info)		1,950
Building maintenance & construction		3,500
Insurance		1,500
TOTAL		**$93,750**

Anticipated Income:

Lafayette CSA	$30,000
Indianapolis CSA	37,000
Farmers' Markets	15,000
Restaurants	3,500
Food Clubs	7,500
Retail Sales & other	750
TOTAL	**$93,750**

Source: Earthcraft Farm CSA.

EXPENSES

Land rental	$8,850.00
Liability insurance	512.00
Maintenance for driving tractors	2,287.35
Tractor loan payments	5,900.00
Purchase of plow, hay wagon, cultivators	1,606.40
BCS walking tractor; purchase of parts and repairs	894.04
Vegetable seeds, seed potatoes, cover crop seeds, potting soil	3,759.26
Fuel for tractors and trucks	3,037.18
Maintenance for trucks	2,238.50
Reemay (plastic for bugs & frost)	1,281.26
Purchase of cartons & boxes	768.40
Administration (printing and postage)	736.28
Greenhouse materials (flats, peat, etc.)	613.80
Seasonal additional labor	1,116.00
Misc. purchases (lumber, power tools, hand tools, bees, etc.)	2,201.22
NOFA/Mass. organic certification	350.00
TOTAL Expenses	**$36,151.69**

Net Income for the farmers (preliminary figures)—pre-social security and income taxes	
Income for Sue Andersen	13,000.00
Income for Chris Yoder	13,000.00
Income for Tom Cronin	26,648.31
TOTAL net income (for 3 farmers)	**$52,648.31**

Source: Vanguarden CSA.

1995 EXPENSES—1996 PROPOSED BUDGET

EXPENSES	*1995 Proposed*	*1995 Actual*	*1996 Proposed*
Labor			
Full time	$12,000.00	$10,000.00	$12,000.00
Part time	7,099.92	4,050.00	4,000.00
Utilities	2,880.00	2,106.22	2,500.00
Repairs	2,149.92	1,605.74	1,500.00
Seeds	2,325.00	762.46	1,500.00
Supplies	2,793.84	4,321.91	2,800.00
Insurance	1,699.92	1,812.00	1,700.00
Gas	900.00	697.00	900.00
Copies	1,399.92	1,065.72	1,100.00
Taxes	1,173.00	1,163.00	1,180.00
Tools	1,800.00	255.00	1,000.00
Office expenses	649.92	264.00	500.00
Dump truck	39.96	39.96	40.00
Classes, fees, & memberships	1,129.92	1,225.00	1,000.00
Advertising	709.92	689.00	700.00
Misc. expenses	999.96	1,335.00	1,000.00
Fertilizers	499.92	357.00	500.00
Health insurance	3,000.00	2,361.00	3,000.00
Apprentice program	2,569.92	4,607.00	2,500.00
Postage	750.00	608.00	700.00
Employee benefits	399.96	448.40	400.00
Entertainment expenses	19.92	19.92	20.00
Phone	499.92	821.00	500.00
Legal	699.96	.00	700.00
TOTALS	**$48,190.00**	**$40,609.00**	**$41,740.00**

1996 Income Projections:

30 Full Shares at	$550.00	$16,500.00
36 One-Half Shares at	300.00	10,800.00
12 One-Third Shares at	205.00	2,460.00
12 One-Fourth Shares at	155.00	1,860.00
13 Working Member Full Shares at	325.00	4,225.00
1996 Income Total		$35,845.00

Shortfall to be made by Fundraisers (1996 Expenses -1996 Income)	$5,895.00

Source: Happy Heart Farm.

A few CSAs that provide meat, milk, or egg shares as well as vegetables break out their expenses into separate enterprises.

The *CSA Handbook* provides useful work-sheets for calculating both farm expenses and share prices (See the CSA Resources for ordering information).

CALCULATING THE SHARE PRICE

How to set a fair price for shares is one of the most puzzling questions for CSA farmers. The cheap food policy in this country exerts a sharp downward pressure on any attempt to sell food at a price that will sustain farms. I keep discovering different ways to calculate the share price. CSAs that make their budgets public usually divide total expenses by the total number of shares to get the share price. Some CSAs divide shares into various sizes or qualities, such as full and half shares, single-person shares, barter shares, gourmet shares, macrobiotic shares, or winter and summer shares. Jean Mills and Carol Eichelberger divide their season into spring/fall and summer shares and charge more for the summer weeks when the burning heat of Alabama makes production more difficult. Dutchess Farm in Castleton, Vermont, offers more northerly Vermonters a five-week share of vegetables most backyard gardens won't produce until much later in the season. CSAs often price half shares at more than half of a full share because of set costs per share and the additional work involved in making more smaller packages.

Farms which have other markets tend to calculate the share price based on the market value of the produce. Subscription schemes charge a fee based on what the market will bear. At Rose Valley in 1989, I began by adding the farmers' market price of the shares on one date each for early spring, summer, late summer, and fall. I then divided by four to get an average weekly price. Through the first season I used this average price to guide the quantities of vegetables I allocated for the weekly shares. Since the early weeks were lighter, I compensated by adding more food later in the season. Marianne Simmons, a core member who oversaw distribution for seven years, reported that there was occasional grumbling about the price, but only among the wealthiest members. When she pointed out that her teenage son earned about as much for an hour of babysitting as the farm workers, the grumbles subsided.

Some farms set their price based on the average weight of the produce they will give each week. They determine how much they value the produce per pound and then multiply by the number of pounds they intend to provide. At the end of the year, they can then report that members paid only $0.79 or $0.99 per pound for their organic produce all season.

Yet another way of calculating the price is used by a few multi-farm CSAs: unit pricing. The purpose of this approach is to ensure that the various farms are compensated fairly for their contribution to the shares, since some vegetables require more work than others. The farmers agree on a way to convert all produce into a standard unit, for example, a bunch of beets equals a pound of tomatoes equals a pint of strawberries, and assign a value to the standard unit. They then multiply by the number of units to get the share price.

For a few years, Caretaker Farm tried calculating how many mouths they could feed and divided their estimated annual costs by this number to arrive at the charge per adult. Children were thrown in for free. In 1996–97, each adult paid $310 towards the farm's budget of $72,657 (*Farms of Tomorrow Revisited*, p. 178). Sam Smith says that this approach led to misunderstandings, so for 1998 they have gone back to a regular share price.

Jim Rose and Signe Waller of Earthcraft Farm CSA in Indiana have a novel system. Instead of a share, the farm sells members a cluster of 180 units which they can use in any combination they choose over the course of the season. Members allocate their units as they wish—a dozen one week, none the next—or give a few to a friend. The farm's weekly menu assigns unit values to the produce. A unit might represent a pound of beans, two pounds of collards, a bunch of green onions, a pint of raspberries, or one pac choi, depending on the supply available. As Signe explained to me, this system works only for the section of their CSA that picks up their shares at the Earthcraft farmers' market stall. For the Indianapolis members, the farm packs the usual weekly bags. Signe says they feel real differences in member responses; the more-local members are more satisfied, loyal, and likely to stay with the project.

In 1995, Tim Laird found that the average price for a full share was $324, with a range of $100 at the low end to $735 at the top. Nineteen of the sixty-five CSAs offered half shares, with an average price of $244. Most of these farms charged more than half the price of the full share to cover set costs per share, such as bookkeeping, postage, and other overhead. Intervale Community Farm and Quail Hill Community Farm charged an extra-high fee for half shares in part to discourage people from buying them. In the 1995

In 1995, Tim Laird found that the average price for a full share was $324, with a range of $100 at the low end to $735 at the top. Nineteen of the sixty-five CSAs offered half shares, with an average price of $244.

season, Quail Hill eliminated half shares altogether and settled on a single-share size.

Member participation can influence the price of shares. Some CSAs offer members the opportunity to reduce their share price by working on the farm. The going value of members' time runs from $5 to $6 an hour. Meeting Place Farm in Ontario, Canada, cuts $50 off the full-share price for 10 hours of work and $25 off the half-share price for five hours. New Town Farms also cuts its price by $5 per hour of work, but asks for twenty hours. Salt Creek Farm in Washington gets away with only $3.50 an hour. GVOCSA keeps share prices lower by requiring that all members work either on the farm or in administration of the project. I calculated one year that if we had to pay for all the work our members do, we would have to raise the share price by $200. (See chapter 8 for more on members at work.)

To attract lower-income members, a few CSAs ask for payments on a sliding scale. Rising River Farm CSA in Washington has a sliding scale of $425 to $500 for a full share and $212.50 to $250 for a split share. The GVOCSA began with a scale of $5 to $7 a week and gradually went up to $10 to $16. Members decided for themselves how much they could afford to pay, with no attempt to verify their income. On the CSA brochure and at the orientation session, core members explained that the actual value per week was $11, then $12, then $12.50, and asked members to pay at least that, if they could. People who paid more understood that they were subsidizing those who paid less. For eight years, by some miracle, the higher payments exactly balanced out the lower payments. Then in 1997, the average came to only $11.92, resulting in a shortfall in payments to the farm. Due to a combination of complacency and distraction, no one noticed the discrepancy until the end of the season, when it was too late to make adjustments. A newsletter plea to members brought in enough voluntary contributions to make up the shortfall.

Rising River Farm
Rochester, Washington

Scholarship funds are another way to provide for low-income members. Many CSAs solicit contributions from members on their annual commitment forms. Vanguarden CSA in Massachusetts explains their request: "We at Vanguarden are serious about ensuring that no one is denied fresh, healthy produce due to financial constraints. Your contributions make this possible. Thanks." The Rochester Downtown United Presbyterian Church, to which some GVOCSA members belong, has made several generous contributions to the scholarship fund. A portion of the price of every *FoodBook* (the Rose Valley Farm crop description and recipe book) has also gone for scholarships, so that in 1997 ten members paid as little as $3 to $7 a week. Some of the payments were in food stamps. The difference between the payments and the average cost of $12.50 a week came out of the scholarship fund. (See chapter 10 for more on food stamps, and chapter 20 for more on involving low-income people.)

To entice early sign-up or enlist member help in recruiting others, some CSAs give a discount on the share fee. If you are considering this, you might ask some of your members whether they need these "bonuses." You may find they are willing to sign up earlier or recruit friends without a financial incentive. The core group of the GVOCSA decided against a discount for members who paid in full before the season because they felt it was unfair, awarding a financial break to the people who needed it the least. Angelic Organics gives a $100 discount, and Full Belly Farm in California gives a free share to members who are willing to host a distribution pickup site in their garage or on their back porch.

In 1995, Gerry Cohn did a study of consumer motivations in joining CSAs in California. With the

MARILYN ANDERSON

Genesee Valley Organic CSA members harvesting Chinese cabbage at Rose Valley Farm.

help of four CSA farmers, he created a four-page survey with questions about why people join, preferences on share size and pricing, alternative sources of produce, level of interaction with the farm, core group participation, and household demographics. He received 235 responses from members of ten CSAs. He found that most of the members had fairly high incomes and education levels. Price was not an important question in deciding whether to join a CSA. Of those members surveyed, 64 percent were willing to make donations towards scholarships for low-income shares (Cohn, "Community Supported Agriculture").

Researchers Jack Cooley and Daniel Lass set out to determine whether membership in a CSA actually saved consumers money. In their survey of 250 members of four CSAs in the Amherst area of western Massachusetts, they found that nearly half believed that they were paying about the same amount of money or more than they would for the same produce in a local store. To test if this were true, Jack found out what was in the shares each week and then priced those items in three different local groceries: a national food chain, a regional food chain, and a local store. Only the regional food chain carried organic produce. Out of the eighty items, fourteen could not be found in the stores at all, so the researchers estimated a price by multiplying the conventional price by 1.5122, which they found to be the ratio of the average price for all organic items to the average price for all the conventionally grown items. They did not include herbs and flowers in their cost comparisons.

Two of the farms charged $450 for a full share; one provided 24.7 pounds of produce per week and the other 27 pounds. Both of these farms used machines for planting and cultivation and irrigated a large portion of their fields. A third farm, a smaller operation with no mechanization or irrigation, provided 8.5 pounds a week for $250. When compared to the retail price for organic produce, the members of the first farm saved $548, of the second saved $682, and of the smaller CSA saved $149. Even compared to the price for conventional produce, the CSA prices were cheaper (Dan Lass and Jack Cooley, "What's Your Share Worth?"). A cost-conscious member of the GVOCSA did a similar comparison with similar results.

A few farms are using this kind of comparative market information in advertising the benefits of joining their CSA. The Cate Farm brochure boasts that members will receive a $700 value for only $450. Dutchess Farm touts a money-saving value of $430 worth of produce for only $320.

In his preliminary results from the first year of a three-year study of CSA economics in the Northeast, Daniel Lass draws a different conclusion: he suggests that farms could be charging more for what they provide. His research reveals that many farms are undercharging for their produce and underpaying their farmers.

Only members who actually make use of all or most of their share truly appreciate the competitive value. For those who don't like to cook or who eat out often, the vegetables may amount to expensive compost. Before the Lass-Cooley survey, almost half of the respondents did not realize that they were saving money. This perception may be a result of paying a large sum for the share all at once instead of paying in dribs and drabs as most people are used to when they shop. One year, a lawyer member of the GVOCSA made a loud demonstration at the signup meeting of paying for his entire share upfront. He insisted that I be summoned to witness his largess. To my astonishment, I discovered that he was paying at the bottom of the sliding scale. When I laughingly pointed out to him that the share's actual value was $3.50 higher a week than what he was paying, he promised to make up the difference. People are funny about money.

BOOKKEEPING

Using a computer program like Quicken or Quick Books may make bookkeeping easier, though the electronically impaired might not agree. For the GVOCSA, two members of the core group share the bookkeeping job: Dennis Lehmann keeps meticulous records by hand, Judy Emerson uses a Lotus Spreadsheet computer program, and at the end of each season they compare the results. They keep track of member payments, dues, and administrative expenses. I have kept the books for the farm as a separate business in a big, old ledger using IRS headings. (See chapter 11 for other kinds of records farms should keep.)

Jill Agnew has a simple way of accounting for member payments. At the farm, she keeps an index box with a card for each member. The members are on the honor system to keep track of their own payments and pay at whatever interval is convenient. When she receives checks, Jill records them on a receipt pad and thus has a double-entry system. She says she has never had anyone who failed to pay.

Harmony Valley Farm in Wisconsin has been able to arrange with its bank to have automatic monthly transfers from members' bank accounts. The annual CSA contract includes an authorization form for the transfer.

CONTRACTS

The CSA brochure or the annual contract or commitment form is where the members first encounter the price of their share. The typical contract has spaces for name, address, and phone number; weekly or seasonal fee for the share and for any additional shares, such as eggs or flowers; preferred pickup day and site; and signature and date. Some contracts also reinforce the message that joining a CSA means agreeing to share the risk with the farmers. The contract of Common Ground CSA in Washington spells out this mutual obligation in detail:

> I understand that I am making a commitment to Common Ground Community Supported Agriculture and recognize that there is no guarantee on the exact amount of produce I will receive for my share. The CSA harvest is protected by sustainable and organic practices such as organic soil improvements, crop rotation, and irrigation. I will share both the rewards and the risks of the growing season along with the other members and the growers. It is my responsibility to pick up my share within scheduled pickup hours. If it is not picked up, I understand that it may be donated elsewhere.

STARTUP EXPENSES

"If you want to make a small fortune in farming, start with a large one," suggests traditional farmer wisdom. Precisely how much you need to start up a CSA depends on a multitude of factors: the buildings already in place on or near the ground you plan to farm; the level of intensity of the methods you plan to use; whether you will need irrigation; and the amount of volunteer labor available. Estimates range from $11,000 to $250,000. Choosing the level of technology that is appropriate for you, your skills, and preferred lifestyle is one of the most important decisions you have to make.

The $11,000 figure comes from *The Rebirth of the Small Family Farm: A Handbook for Starting a Successful Organic Farm Based on the Community Supported Agriculture Concept,* by Bob and Bonnie Gregson. Anyone thinking about entering farming should read this charming little book of practical pointers based on the Gregsons' own lives. They list the equipment and materials you would need for a two-acre market garden where most of the work is done by hand. The income they estimate is from sales of gourmet produce to middle- or upper-class customers.

CSA Works, Michael Docter, Linda Hildebrand, and Dan Kaplan's consulting firm, suggests an investment of $15,000 to $20,000 for the farm implements you would need to cultivate five acres or more using mechanical weed control systems. They recommend the purchase of an Allis Chalmers "G" tractor with cultivating implements and a Planet Junior seeder. They offer help figuring out if you are ready to start a farm or, if you are farming already, how to make your operation more efficient. (See the CSA

STARTUP COSTS FOR A TWO-ACRE FARM

Item	*Rough Cost*
Rototillers	$2,900
String trimmer	350
Hand tools	200+
Carts and wheelbarrows	500
Temporary shelters	700
Perimeter fencing, 1/2-acre	600
Cold storage	900
Irrigation equipment	700
Ground cloth	400
Raised growing beds	200
Metal tee posts & netting	400
Seeds	300
Washing/mixing tubs	100
Scales	100
Chicken housing/fencing	300
Miscellaneous hoses, materials, etc.	800
Soil test & amendments	300
Business license	100

Source: Gregson, *The Rebirth of the Small Family Farm.*

Resources section for more details.) The $20,000 figure does not include buildings or the larger tractor a farm would need if you were planning to do your own plowing and make compost on a large scale. While exemplary in many ways, the Food Bank Farm is not self-sufficient in terms of on-farm fertility; if that is one of your goals, your equipment costs will be higher.

Jean-Paul Courtens sketched out an equipment list and rough costs for the level of capitalization in place at Roxbury Farm. Not including the land or livestock, he estimates an investment of $250,000 for barns, greenhouses with a vacuum seeder or soil block machine and flats, three tractors, a manure spreader, tractor implements for tillage, cultivation, transplanting, seeding, and haying, a harvest truck or flatbed wagon, a delivery truck, a root-crop washer, irrigation equipment, a walk-in cooler, a bucket loader, and a forklift. I think Jean-Paul was pricing new equipment. At Rose Valley Farm we had a comparable level of capitalization, but by doing the construction ourselves and purchasing mainly used implements, we kept the costs at less than one-quarter of Jean-Paul's estimate. Like many other small farms, we also made our purchases gradually, adding one or two key pieces of equipment each year as we could afford them out of current earnings.

Bill Brammer of Be Wise Ranch has expanded his operation every year for twenty-one years, starting from three-quarters of an acre and a rototiller, to over three hundred acres with several tractors. His latest acquisition is a machine that enables him to roll up and reuse drip irrigation tape five or six times. By combining drip irrigation with transplants, Bill has reduced water use from 3 acre-feet to 1 acre-foot for some crops, a major saving when water costs $640 per acre-foot. One of the advantages of getting larger, Bill says, is that the increased volume of sales has provided the capital to purchase adequate equipment. Although he depends on compost to improve the rocky, clay soils mainly used by other farms for cattle ranching, he only recently purchased a manure spreader to mechanize the job.

To make good use of equipment demands skill, experience, and careful planning. The more experienced you become as a farmer, the more ways you discover to make your farm more efficient, more conserving of energy (both purchased and human), and more closely in harmony with the natural forces

RECOMMENDED YEAR-ONE EQUIPMENT LIST FOR A FIVE-PLUS-ACRE STARTUP CSA

Item	Cost
Medium Tractor, 35–40 HP, 3-point hitch, PTO, narrow tires	$4–6,000
Disc Harrow with heavy drag bar for seed bed	$5–700
Buddingh Rotary Hoe for stale bed preparation, single-row cultivation (for two-row cultivation in future years)	$5–600
Planet Junior Seeder with standard shoe and scatter shoe	$250–350
Drop Fertilizer Spreader (lime spreader) for granular fertilizer, 7–10'	$2–300
Flame Weeder	$150
Farm pickup truck	$300–1,000
Hand tools, hoes	$150
Irrigation pipe (site specific)	$2–2,500
Irrigation pump	$2–4,000
Allis Chalmers "G" tractor with one row, hilling shoes, rear sweeps and/or Farm-All Cub tractor with one-row cultivator	$2–3,000
Potato equipment	
Planter	$400
Digger	$300
Spray rig	$700

Note: Prices assume used equipment. Much of this equipment is obsolete by today's farming standards, but is perfectly suited for diversified CSA vegetable production. Due to its obsolescence, it is inexpensive.

Source: CSA Works.

Would You Please
Add a Little Glitter to that Squeaky Ripper?
(John Ponders Metal and Muscle)

We have thirty-three machines on this farm. They have names like discer, hiller, harrow, digger, mulch layer, field cultivator, cultipacker, barrel washer. The machines aerate, pulverize, plant, scrub, dry, hill, till, weed, rip, mix, smooth, cool, shuttle. They have shanks, knives, bearings, teeth, scrapers, brushes, lathes, belts, wheels, springs that bend, break, screech, corrode, wear in, wear down, wear out. They need greasing, sharpening, inflating, welding, patching, riveting, tightening, drilling, shimming, propping, oiling, jerry-rigging.

This is a little farm, and that's a lot of machinery propping it up, and occasionally dragging it down. Most of it is used. Some is ancient, like our Iron Age (actual brand name) potato planter—originally designed for hitching to a horse. Much is cast off from farms that specialized and expanded, and some is residue from farms that folded.

The vast majority of work here as measured in hours is through hands-on physical labor. I keep looking for "labor-saving" devices to create more balance between technology and the human back. I read the classifieds and watch the sale bills. I drive to Michigan and up into Wisconsin in the hopes of locating obscure specialized equipment that will ease the sheer physical demands of bringing a weekly box of produce to your neighborhood from some packets of seeds the previous winter. Imagine all the different actions involved in that metamorphosis—procuring the ingredients for the greenhouse soil, mixing, sieving, planting, watering; preparing the fields: composting, tilling, transplanting; tending the crops: weeding, irrigating, thinning, foliar feeding; harvesting: discerning, cutting, picking, digging, peeling, separating, hanging, drying, trimming, snapping, bunching, lugging, conveying, washing, sorting, divvying.

I have an affinity for machinery. One machine can sometimes do the work of twenty laborers. It will never sneak out of the field in the middle of the morning for a nap. It's usually where I left it. If it breaks I can almost always fix it. And I've never seen a drunk chisel plow.

On the other hand, the 656 Farmall tractor has never made me lunch, I've never flirted with the wheel hoe, and I've never enjoyed an Amazake White Russian with the tiller.

(I was going to delete these last two paragraphs, but Jill and Cindy insisted that I leave them. Editing talent is another quality that my machinery lacks.)

Fieldhands have other redeeming qualities that give them a special status here. They plant and water the seedlings, clean the packing room, yank weeds, discern ripeness, maintain machinery, stock our stand, and fill and deliver your boxes week after week with vegetables they harvested, cleaned, and graded almost totally by hand.

And as I learned Saturday night, the workers dress splendidly for a farm party and laugh 'till their sequins drop. My tillage tools reposed obediently, drably, in the full moonlight of the farmyard, as the farmhelp spun, glittered, and danced.

—John Peterson, Angelic Organics, Caledonia, Illinois. Used by permission.

on and around your farm. A workshop at the Upper Midwest Organic Conference in 1998 compared the economic "facts" and "ratios" of two CSA farms. Earthcraft Farm added a column with their own figures and sent me a copy. They compared numbers of shares, acres, gross and net farm income, farmer person days, total person days, and payroll expenses. The resulting ratios were very ambiguous. I think we can learn from one another, but comparing farm statistics may not be the best way. Training in Holistic Management offers a powerful set of tools for improving decision-making on any farm. Organizations such as NOFA-NY have just begun exploring the possibilities of mentoring, pairing new farmers with farmers who have more experience. Workshops and conferences where CSA farmers and active members share open-handedly with one another are of tremendous value. We can all run our farms so much better if we pool our best discoveries.

Despite CSAs' lofty goals, there is no evidence yet that they *guarantee* a comfortable income for farmers. Two small studies found that about half the farmers reported increased incomes when switching from previous marketing arrangements to CSA. For nearly as many, income remained the same. But almost all the farmers felt less insecure and expressed optimism that their incomes, though still low, would grow as the CSAs matured (Laird, p. 78–80; and Kelvin, p. 20). This hope is borne out by the farms profiled in *Farms of Tomorrow Revisited*, by Trauger Groh and Steven McFadden: While not making out like bandits or bankers, after seven more years of development, the farmers were doing better financially.

In 1993, Jack Kittredge interviewed a farmer with a twenty-five-member CSA who shared these reflections on his farm's economics:

> Small scale makes sense ecologically. I wish there were more CSAs. A local one went out of business this year because he owed too much money for land, mortgage, and equipment. A successful CSA needs to grow cautiously. It needs to be more of a lifestyle choice than a way to make money. I couldn't possibly do this if I owed money. We own our land and house with no mortgage. Taxes are less than $500 a year. I haven't figured out how most farmers survive, especially if they want to live like what I perceive to be the "average American." ("CSAs in the Northeast," p. 12)

◆

We must not expect to divorce CSA farms from the world context in which they exist. As Jim Rose and Signe Waller said in their "Five Year Retrospect and Prospect": "The entire food system, from the most personal level of the shopping and eating habits of advertising-hounded consumers to the macroeconomic level where world food prices are controlled by a few corporations, constitutes an environment hostile to small and middle-sized farmers. It is almost a mission of insanity to undertake to farm as more than a hobby, to actually try to make a living at farming." Miraculously the CSA connection is helping farms to do just that!

Legalities 10

Character is the internalization of responsibility. What we are talking about when we talk about a local food system or CSA is a food system that relies more on character than it does on legal, bureaucratic, or commercial procedures.

—Wendell Berry,
quoted in *Safe Food News*

In this book we can offer only very general information on the legal aspects and implications of CSA. Laws vary from state to state and county to county. You may need to consult a local lawyer. Better yet, recruit one as a sharer and barter for the advice. This chapter will at least give you enough information to formulate the right questions.

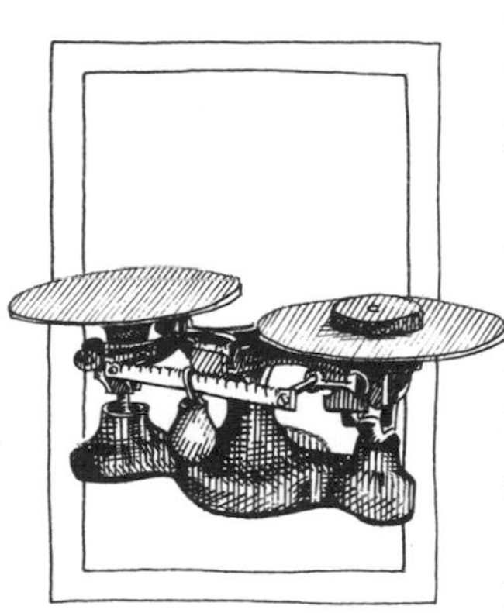

INSURANCE

Most CSAs carry standard liability insurance. As separate coverage, liability can be very expensive; as part of a farm insurance package, the price is more reasonable. You should try to get a liability policy that includes a stated level of medical expenses paid out without a lawsuit. Some CSAs have additional liability as a special form of "pick-your-own." The rates for "pick-your-own" will be lower if you specify that you do not use synthetic pesticides and members do not use equipment, horses, or ladders. Take the time to explain to your insurance agent how your CSA differs from a regular pick-your-own operation, and provide documentation, such as a copy of your brochure or newsletter. If an accident should occur, your agent will not then be able to plead ignorance about the nature of your operation.

Pick-your-own coverage will allow members to help harvest and to use hand tools. Keep a first-aid kit handy. Encourage members who work regularly to get a tetanus shot to reduce the danger from puncture wounds.

If children are welcome at your farm, you must have clear farm rules. Unless you are providing childcare, parents or guardians must understand that they are responsible for their children. At Rose Valley Farm, when children arrived, we rounded them up, got their attention, and told them our three rules:

1. Children are not allowed in the sheds or barns or on any farm equipment unless one of the farmers is with them;

2. They must learn to recognize poison ivy and stay out of it, but, if they think they have touched it, tell one of the farmers, who will help them wash off the toxin; and (they all look tense and serious by this time),
3. Have fun! (laugh of relief)

Silver Creek Farm has a pond where farmers and members can swim. Molly Bartlett makes sure all visitors know that no one is allowed to swim alone, that children must be supervised, and that there are set times when everyone can swim. (See "Children on the Farm," p. 86 for more on this topic.)

In this machine age, children love to ride on farm equipment. Satisfying this desire is tremendously seductive. It makes happy children who express themselves in shrieks of delight (if you give them a good rough ride) and warm gratitude. Enter the insurance grinch—no affordable policy exists that would protect your farm should an accident occur.

If you are going to welcome outsiders to your farm, you must make every effort to reduce hazards. Clean up junk piles, pick up rusty nails, old tines, pieces of equipment, and anything lying around that can poke, pinch, or cut your visitors. Install a few warning signs, such as "Employees Only" on the equipment or tool shed. Distribute a map of the farm to CSA members with restricted areas clearly indicated. An insurance agent Robyn consulted said, "forewarning members is your best protection."

HEALTH INSURANCE

When you succeed in creating agriculture that is truly community-supported, you won't need health insurance. Should a serious health problem occur, your community will bring forth the contacts or funds you need. Unfortunately, most of us have a ways to go before we reach this level. In the meantime, the CSA must decide whether to factor health insurance for the farmers into the annual budget or leave it up to them to pay for themselves.

The legal structure of the CSA will determine how health insurance is handled. If your CSA is a sole proprietorship or a partnership, the farmer or farmers will have to purchase insurance for the self-employed. If the CSA incorporates, either as a for-profit or nonprofit corporation, and hires the farmer as an employee, it should include health insurance as part of the benefits package. If the farmer is an independent contractor, responsibility falls back on the individual.

FOOD STAMPS

Before they became CSAs, many farms that sold at farmers' markets or through farm stands had received authorization to accept food stamps in payment for produce. When they became CSAs, they continued accepting food stamps without problems. However, the USDA has turned down a few CSAs that had included a full explanation of CSA with their application for food stamp authorization. In a 1994 policy memo, USDA stated that CSAs are not eligible on the basis of collecting payment up front at the beginning of the season. CSAs that are nonprofits and deliver the food within two weeks of receiving payment remain eligible for food stamps, as are CSAs that operate like farm stands.

The National Campaign for Sustainable Agriculture and the Community Food Security Coalition have taken issue with USDA over this ruling. Conceivably, with enough public pressure, USDA will be persuaded to change its regulations or Congress will change the statute. CSAs can wait for the rules to change, or they can apply to USDA as market gardens or farms with on-farm sales to customers and fit in under this familiar category.

Meanwhile, the Electronic Benefits Transfer (EBT) threatens to pose an even greater risk to small farm acceptance of food stamps. Instead of issuing food stamps, USDA will give recipients a plastic card, much like a credit card, that will register when passed through the appropriate electronic equipment. For major food chains, which already have such receptors, EBT will entail a minor retooling. For small farms, purchase of this equipment would be prohibitively expensive. This issue, too, is on the agenda of the national sustainable food networks.

To apply for food stamp authorization, call or write your nearest regional USDA office for an application form. They usually take a few weeks to process your application. Once you receive authorization, you must apply to your bank for an official rubber stamp with your number on it, which can take an additional six weeks. The bank will not exchange Food Stamps for cash, but will require you to deposit them in an account. USDA will probably invite you to attend a training session on the proper uses of food stamps.

The reauthorization process that USDA requires from time to time is more elaborate than the original application. Reauthorization forms provide no category that fits a fresh market farm. The local representative responded to my queries by advising me to fill the forms out as best I could. Many moons later, although I have still heard nothing, the bank continues to accept my food stamp deposits.

Once you receive authorization, include the food stamp payment option in advertising your CSA to encourage participation by lower-income people. To accommodate lower-income members, you will also have to allow more frequent payments. Food stamps are issued monthly, and the regulations stipulate that recipients can make advance purchase transactions at two-week intervals only.

CSA LEGAL STRUCTURES

A CSA can adopt a variety of legal structures. Each group should determine which form is most appropriate. Some CSAs are "sole proprietorships" or partnerships; in other words, both farm and CSA business are the property of the farmers. Other CSAs separate the CSA from the ownership of the land. The land may be held as a sole proprietorship, a partnership, or a corporation, while the CSA is an unincorporated association or is incorporated as a nonprofit corporation. Groups of farmers can organize as farmer-owned cooperatives, most of which are corporations. There is no set structure in the law for food co-ops or buying clubs, so groups of consumers can choose the corporate structure that suits them best in forming a CSA. Institutional CSAs usually hold both the land and the CSA as part of a nonprofit corporation. Each form has advantages and disadvantages. The details of these legalities will vary from state to state.

Robyn Van En sketched out the following information based on Massachusetts state law:

Sole Proprietorship

- Not regulated by the State Statute; consequently no legal help is necessary to establish it.
- Proprietor is responsible or liable for debts and obligations of the business or property.
- Proprietorship income can be shown on a schedule of the proprietor's individual tax returns.
- Proprietor pays half of Social Security and Medicare, employees pay other half.
- Basic bookkeeping is necessary.

Partnership

- Similar to sole proprietorship, but requires registration of partnership with county clerk. In forming a partnership, it is wise to write a partnership agreement, which includes how the partnership will be dissolved. *The Partnership Book* (Berkeley, Calif.: Nolo Press, 1995) contains all the clauses and instructions on how to form your own partnership.

Limited Liability Partnership

- General partner runs the business, and has rights and liabilities of any partnership.
- Limited partners have no management powers, and are liable only to the limit of how much they have invested. Limited partners are basically investors.

Unincorporated Association

- Not regulated by the state, no filing fees necessary.
- Administering individuals are responsible and personally liable for obligations and debts.
- All labor may be contract labor, and contractors are responsible for own taxes and insurance.

Incorporation

- Filing fees are necessary (usually about $500, plus possible legal fees to process the paperwork). All activity is governed by the state.
- Creates a legal entity, separate from stockholders, that can hold a mortgage or long-term lease.
- Legal suits would be directed against the assets of the corporation rather than individuals. The owners (stockholders) are not personally liable.
- Separate books must be kept and separate tax returns must be filed.
- A Board of Directors must be elected or selected to manage the administration of the corporation, which might include hiring and firing of employees.
- Stockholders and shareholders invest with the expectation of a financial return on their investment.
- Takes tax deduction for compensation paid to employees, including shareholders who are employees.

Farmer-Owned Cooperative Corporation

The Capper-Volstead Act of 1922, a Federal law, authorized the formation of farmer cooperatives. The USDA Rural Development office (Stop 3257, 1400 Independence Ave., SW, Washington, D.C. 20250; 202-720-8381) will send a packet of information on how to form a farmer cooperative with sample legal documents, policies, and a history of co-ops in the United States.

- The structure and liability limits are similar to a corporation as described above. The minimum number of farms required for incorporation varies under the statutes of different states. Most states require that a cooperative have officers, a name, an annual membership meeting, and a mailing address.
- Five underlying principles distinguish cooperatives from other types of private businesses:

1. Ownership is by member-users.
2. Control is on the basis of one vote per member, or on volume provided.
3. Operations have an at-cost (nonprofit) objective.
4. Dividends on member capital are limited.
5. Education is necessary for understanding and support.*

Nonprofit Corporation

- The procedure for organizing a nonprofit is very similar to forming a corporation, including the filing of fees. An important difference is the ability of nonprofits to receive grant funding.
- Investors do not expect financial return on their investment. There is no capital stock or stockholders.
- Nonprofit tax status is required to be eligible for charitable donations, which are tax-deductible to donors.
- Assets can be transferred only to another nonprofit should the organization cease operation.
- If the nonprofit qualifies for 501(c)(3) status, all activity is exempt from federal and state taxes.

* From "Cooperative Principles and Legal Foundations," Cooperative Information Report 1, 1977, USDA Agricultural Cooperative Service.

The unincorporated association or sole proprietorship forms are the most appropriate for the first year or so while a CSA gets itself together. Many farms, while remaining under the operation of a family, incorporate to limit liability.

The members of the Hudson-Mohawk CSA invited Janet Britt to be their farmer, and then chose to classify their CSA as her sole proprietorship. In an article in *Harvest Times* (Winter 1992), Janet explains the decision: "The people who had come together to get our farm started were willing to work hard to organize it, and to have the trust and faith to buy a share, but were not ready to assume legal liability for the project. I was willing to assume responsibility because I wanted our CSA to start and had few personal assets to lose in the unlikely event of legal problems." While legally a sole proprietorship, the group agreed to a philosophy of cooperative ownership, which meant that Janet owned all capital investments, but agreed to leave these assets with the CSA if she were to leave.

Rose Valley Farm, which produced most of the food for the GVOCSA for nine years, was the property of David Stern and myself, and we ran our farm business as a partnership. The GVOCSA, however, is a branch of the Politics of Food, a nonprofit 501(c)(3), and does not make any profit. It passes all the money it collects in food payments to the farm. An annual membership fee covers administrative costs. From the point of view of the various authorities, the GVOCSA contracted with the farm to purchase the crop and was responsible for picking and distribution. When members worked at the farm, they were not working as farm employees. I did not have

Sharon Hamer checks her name off at the Genesee Valley Organic CSA pickup center at the Lutheran Church of Peace in Rochester, New York.

to fill out 160 W-2 forms. The farm and all farm equipment belonged to the farmers, but the cooler used for distribution belongs to the GVOCSA.

Qualifying for nonprofit status can be tricky. State and federal regulations govern the kinds of organizations that qualify and the activities that are acceptable. Failure to make a profit does not in itself qualify a farm to be a nonprofit corporation. An established nonprofit can more easily include a CSA among its activities than a new CSA can set up its own nonprofit. Donations to a nonprofit must be spent on acceptable activities such as educational programs, and religious or scientific activities open to the general public. A CSA whose primary goal is to feed and educate its own members would have a hard time qualifying as a 501(c)(3): the benefits have to go to society at large, not to a limited group. A nonprofit that adopts a CSA as fitting with its mission might serve as the funds administrator for the CSA and require a small fee for the service. It might also request that the CSA handle some of the required paperwork. The GVOCSA fits very well with the community food security work and urban-rural linking done by the Politics of Food.

Robyn concluded her notes on CSA legalities with some reflections on future growth and possible threats to the spirit of CSA: "As I write, CSA is still a little-known alternative, but the concept has tremendous potential to engage masses of people. Genuine CSA, where members have opportunities to meet the farmers and production crew, will be difficult for agribusiness to co-opt. But like the word 'natural,' which we see plastered all over many things, natural and otherwise, the concept is vulnerable to the more money-minded to put their spin on it. Increasing the scale of CSAs will be necessary to feed the millions of people in urban centers. The challenge is to retain the integrity, quality, and underlying philosophy, as well as the community-building functions. I believe the answers will come; it is all unfolding as we go. In its grossest terms, CSA is still just a marketing technique. Each community is responsible to develop and perpetuate the more subtle values and benefits that are near and dear to many of us."

TO CERTIFY OR NOT TO CERTIFY? 11

◆ *To produce food of high quality.*
◆ *To interact in a constructive and life-enhancing way with natural systems and cycles.*
◆ *To consider the wider social and ecological impact of the organic production and processing system.*
◆ *To encourage and enhance biological cycles within the farming system, involving micro-organisms, soil flora and fauna, plants and animals.*

—FROM "THE PRINCIPLE AIMS OF ORGANIC PRODUCTION AND PROCESSING,"
IFOAM BASIC STANDARDS

WHILE ORGANIC CERTIFICATION is not a legal necessity for a CSA, if your farm is organic or Biodynamic, you should consider it carefully for several reasons. Certification will help recruit new members who are seeking organically grown food, and do not yet know the CSA farmer personally. At its best, the certification process can be the occasion for an annual reflection on your farming practices and progress in building soil health and biodiversity, as well as for thorough record keeping. Farmer- or farmer/consumer-run associations sponsor many certification programs, so membership brings access to lively and informative newsletters and farmer networks. The negative aspects can be the cost and the paperwork. Some farmers do not want to be regulated, even if they agree with the regulations.

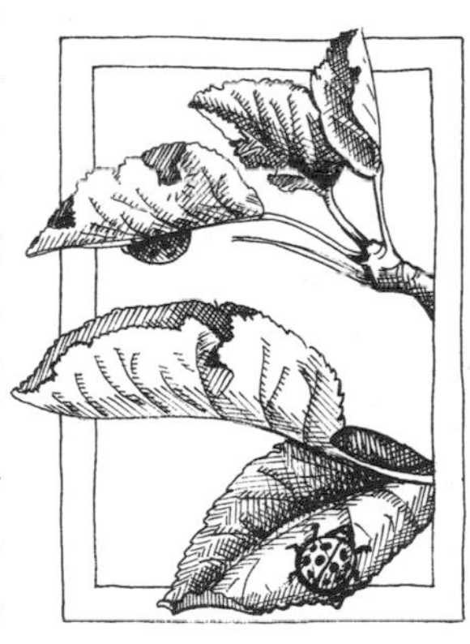

The Organic Food Production Act (OFPA), the federal organic legislation passed in 1990, if implemented, will require that any food to be labeled "organic" come from a farm that is certified by a USDA-accredited certification program. Should a CSA decide not to certify, it will not be legally permitted to advertise itself as organic. The members may not care—knowing and trusting the growers may be all the assurance they need. But the lack of the organic label may make recruiting new members more difficult.

A move is afoot to exempt CSAs from the mandatory requirements of the OFPA. At the time of this writing, there is no way to predict the results. However, a problem that comes to my mind is the small percentage of CSA farms that market exclusively to their membership. For instance, of the thirty CSAs in Vermont, only one is a "pure" CSA. All the others have outside markets. Fairness would seem to dictate that any direct-marketing farm should have the same exemption. My bias is to push for a change in the legislation to remove the requirement that certification be mandatory. Let the National Organic Program control the label "certified organic," and allow farmers to make their own decision about whether to certify or not. The entire system of verifications, in any case, is only as good as the integrity of the farmers.

Under the OFPA, the basic certification document will be an organic farm plan, which may or may not coincide with the annual application form. Created by the farmer, the plan will describe soil and livestock management, pest and disease control, and use of water resources, and will set out goals for continual improvement in these areas.

Depending on where you farm, certification can be an expensive proposition. Private programs charge a basic fee, inspection expenses, and a percentage of sales. Tax dollars support a few state-run programs. A CSA may decide that the expense is too high for the farm budget, especially in the first years.

Redwing Farm
Townshend, Vermont

Tim Laird writes that organic certification was an issue for both of the CSAs for which he farmed. The Intervale Community Farm considered certification crucial for recruiting the fifty or so new members they needed every year. Quail Hill Community Farm decided to certify only after a few years of operation. These are the reasons Laird lists:

- We wanted to support NOFA-NY financially beyond membership dues. As a CSA we added the certification to our budget, so the community as a whole, and with some consciousness, knows that by joining the CSA they are supporting not only our farm, but the movement of organic agriculture as well.
- As farmers, we wanted to feel more a part of the organic farming community. Becoming a certified farm is a natural means to achieve this.
- We wanted to draw on technical support that NOFA-NY offers. (Laird, p. 27–28)

Consider also the value of belonging to a network with other organic farms. In many parts of the country, organic farming associations, such as the NOFAs, MOFGA, Tilth, and the Biodynamic Association, run certification programs, and also serve as educational networks for farmers, providing workshops, conferences, newsletters, and contacts with other growers. In the absence of a supportive land grant university, Extension office, or other government programs, these farming associations have been the primary sources of information on organic farming. CSA growers can benefit from linking with these networks.

Many of the organic certification programs around the country were developed by these associations. Members, both farmers and consumers, negotiated practical standards for farmers that meet consumers' needs for protection in the marketplace. Many hundreds of volunteer hours, the conscientious efforts of farmers and consumers thinking together, have gone into the formulation of organic standards, including both the International Federation of Organic Agricultural Movement (IFOAM) and the more recent federal standards embodied in the OFPA. The IFOAM standards, dating back to the 1970s, served as guidelines for most certification programs all over the world. The wording of the OFPA section on materials, which came from the members of the Organic Farmers Associations Council (OFAC) and the work of Lynn Coody in particular, articulates the experience of certification programs in their struggle to develop criteria for judging what kind of materials have a rightful place in organic agriculture. If respected in the implementation of the National Organic Program (NOP), these criteria will ensure that accredited organic certification programs will approve only ecologically sound materials. The Organic Materials Review Institute (OMRI), recently established with the support of most of the existing private certification programs, uses these criteria as the basis for its materials list.

Seven Oaks promotes their Vermont state organic certification.

When CSA growers use the standards of a certification program to guide their growing practices, they are taking advantage of this enormous volunteer effort. To acknowledge this debt by contributing to the organic movement would seem to fit with the spirit of community supported agriculture. Choosing to be certified is one way. Joining a certification program by becoming a member of a program standards committee is an excellent way to learn more about organic practices. Most certification programs go through an annual review: participation ensures that the particular needs of CSA farms will be taken into account.

Long-term planning and careful record keeping are part of the orderly management of a CSA farm, whether certified or not.

The following records are useful to keep:

- seed and plant purchases
- purchases of soil amendments, pesticides (including botanicals, purchased beneficial insects, pheromones, etc.), farming supplies
- date, amount, and purpose of pesticide applications
- dates of equipment purchases, maintenance, and repairs
- livestock and feed purchases and sales
- date, amount, and purpose of livestock health treatments
- field maps showing rotations, soil amendments, date of plantings, date and amount of harvest
- regular soil tests
- use of water, water tests
- compost-making ingredients, dates of pile creation and use

Because the certification process demands this kind of detailed record keeping, belonging to a certification program can be a useful prod to complete paperwork not all farmers love to do. Certification requires the completion of an annual application form and at least one inspection. To fill out the application, a farmer needs to keep most of the records listed above: records of every input, of field production and yields, and a paper trail that leads from every sale back to every bit of crop harvested.

EACH CSA WILL HAVE TO MAKE its own decision about organic certification, weighing the advantages and disadvantages. When we reach that ideal state in the marketplace when the label on every food item details where and how it was produced, all the materials used by the producers, and how much energy was expended in production, and when every buyer is sophisticated enough to read and understand those labels, we won't need certification anymore. We have a way to go yet. . . .

COMMUNITY AND COMMUNICATIONS 12

Community is formed by people who are acting in cooperation with one another.

—JEANNETTE ARMSTRONG, "SHARING ONE SKIN."

THE BRIEF HISTORY OF CSA in this country dates back only to Indian Line Farm's Apple Shares in 1985. We are still in the pioneering phase—anyone who ventures onto this frontier should expect to share the risk. CSAs that are thirteen, eleven, or even nine years old are the venerated veterans of this movement. What is the glue that holds them together? Despite traumatic upheavals—moves to new pieces of ground, irreconcilable splits within core groups, divorces and financial crises—these CSAs *are* holding together and moving forward. One common trait is their unabashed readiness and capacity to borrow and apply one another's best inventions. Their example and the power of the CSA concept are attracting new people with exciting ideas. What will give these new CSAs the persistance of witchgrass rhizomes or make them spread as irrepressibly as galinsoga weed?

The story of the Genesee Valley Organic CSA is about as tumultuous as any. The project has survived a very serious flood; a moderately severe drought; the estrangement, reconciliation, then definitive separation of the two main farmers; and a move to a new farm. As one of the farmers involved, I appreciate most the way CSA softens the business side of farming. Selling vegetables directly to people whom I know shelters the farm and me from the impersonality and indifference of the global marketplace. Supermarket produce managers I have encountered care more about saving a penny on a head of lettuce than about what happens to my farm. CSA members, on the other hand, write us "vegetable love letters," such as this one from Colleen Fogarty: "How often I think of you farming the land at Rose Valley with awe, admiration, and profound gratitude. At the first distribution this year, I was elated to have fresh greens—lettuce, spinach, arugula with two colors of radishes. It was such a cold spring, it seemed nothing would ever grow, and the miracle of those first vegetables stayed with me many weeks. More words

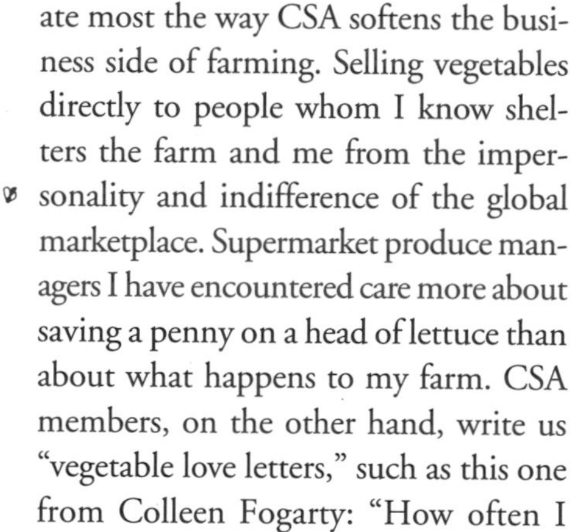

couldn't really convey my thanks to you for coaxing such wonderful food from the ground with such love and care." Such words counterbalance a lot of aching muscles and mosquito bites. The quality of my relationship with the members has moved me to want to stay with the CSA even when I had to leave Rose Valley Farm.

As I peruse end-of-the-year surveys, I find two main attractions have helped keep GVOCSA members: the quality of the vegetables and the farm work. These are some typical quotes from the surveys:

"The quality was excellent! Once I got the hang of it, we ate almost all of it every week and the selection was great. It taught me to cook with things I wouldn't have tried."

"I enjoyed working at the farm—especially the variety of jobs you have to do. I enjoy the warm atmosphere and especially that the children are welcomed."

To the question, "Is the CSA concept worth the trouble?"

"Yes, it is significant to be involved—even a little—with growing food in a healthy way and seeing it through from the ground to the kitchen. I am grateful to have the opportunity. And especially grateful to the core group for their efforts."

To the question, "What did you *dislike* the most?" one member answered: "Getting up early in the morning and working four hours outside." To the question, "What did you *like* the most?" the same

THE TORNADO AT WOLSELEY COMMUNITY GARDEN

On Friday, I experienced my first tornado—not an experience I care to repeat! It touched down on the east side of the garden and took out a swath of eight or so trees. It touched down again in a large field just beyond my yard and flattened a perfectly round area. If you didn't know otherwise, you would think a million-pound space ship had landed there. It touched down again in my neighbor's yard and pulled up a few of her oak trees. Monsoon-type rains accompanied it. It was a complete whiteout, but I couldn't hear the rain because the wind was making jet-plane-landing noises. Of course, you don't get a storm like that without a little bit of HAIL thrown in. There wasn't a great deal of it, but the stones were large.

All this happened in about thirty minutes Friday night. I was sure I was going to wake up Saturday to total devastation. I went to sleep composing a letter to all of you. It went something like this: "Dear Members: I know $250 for one bag of produce is a bit steep but . . . "

On Saturday I had to get up at 5 A.M., load up and head for the St. Norbert Farmers' Market. I could not bring myself to even look at the garden. When I got home I had a power nap and headed over to a friend's house for dinner. When I returned, I couldn't stop myself from going out to the garden. It was a nightmare! Everything was smashed into the ground and covered with mud. I started working on the letter to you again.

Sunday was spent in a depressed haze. I tried not to look at the garden. On Monday, my forty-eighth birthday, I woke up to a gentle rain. It continued for several hours and then the sun came out. I took a walk in the garden and, although the plants had plenty of holes blasted through them, they were standing up again with clean and shining faces.

You will notice a lot of holes in your greens. Just eat around them.

—SANDRA CONWAY,
FROM WOLSLEY COMMUNITY GARDEN NEWSPAGE,
GARDENTON, MANITOBA, JULY 4, 1996.
USED BY PERMISSION.

member responded: "Getting up early in the morning and working four hours outside."

When I asked a dozen of the longest-term members why they have made CSA part of their lives, they always start with the vegetables, but then go on to deeper aspects of the work, the community, agriculture, and food economics.

Marian Vaeth, who joined in 1990, said: "I want to support organic agriculture. I love the produce and the people who are in the program. One year I wasn't a member, when my husband was sick—I missed the produce and I missed the people."

Fred Miller, also a 1990 recruit, had cool things to say about some of the other members, but stayed for the food: "I like the idea of going out on the farm and helping grow the food. I didn't have a place of my own to farm. Farm work is good clean work on an organic farm. No problem with pesticides, herbicides, and I like the farmers too."

Brenda Mueller, a founding member, explained that she has multiple chemical sensitivies and needs to eat natural foods, free of pesticides: "The CSA and you have been flexible, finding work for me that I could do. Also, it is convenient to get all my vegetables in one place, even with the limitations of time."

Eloise Schrag, class of 1990, said: "A big part is my commitment to farmers. Both Dennis and I grew up on farms. Our parents were farmers. Having children, I like the idea of organic vegetables. I grew up gardening and canning and preserving food. I like the quality and the taste. And I like paying the farmers directly."

The GVOCSA is not strong on convenience or service to members. The work requirement has certainly discouraged some people from joining. When new recruits offer to pay more instead of working, we send them to the Porter's CSA across town. Tight parking and a two-hour pickup interval create the most discontent. Especially for people reluctant to ask for help, the necessity of being at the same place at the same hour every week to get the vegetables is a constraint. The core does an annual phone survey of members who quit and tries to make adjustments. These minor inconveniences do not rank high in the reasons for leaving, though one former member gave the pickup time as her reason for switching to another CSA. The main causes of member attrition are:

- moving out of town
- having another baby
- starting own garden
- summer plans—long period of travel or frequent trips out of town
- prefer to choose own menu
- too much food

David Inglis reports that he counts on losing 15 percent of his members every year because of the local population's transience. He added divorce to my list of top reasons. Jennifer Bokaer-Smith's former members gave similar answers. She also had a few members join in order to make major changes in their lives through cleaner eating, who quit when they failed.

David Trumble of Good Earth Farm in Weare, New Hampshire, calculates that over the last six years his eighty-family CSA has had a retention rate of 85 to 90 percent, which means that 50 percent of their customers are still with them, if you accept his math. David stresses the importance of keeping members:

> The issue of retention rate is key to the long-term economic sustainability of a CSA farm. For example, a 90 percent annual retention rate over five years yields a total 54 percent customer retention (90 percent times 90 percent times 90 percent times 90 percent times 90 percent = 54 percent). An 80 percent annual retention rate over five years yields a total 33 percent customer retention. A 70 percent annual retention rate over five years yields only a 17 percent total customer retention. At first, 80 percent sounds pretty good; but if in five years you only have one in three of your original customers, you could be facing long-term problems.

Living Close to the Ground

I am on my knees again. It will be like this until the frost. After we finish the planting, we fall to our knees, pulling weeds and grass, thinning carrots, hilling potatoes, picking peas and beans. Some mornings, I walk down to the garden, and Rose Ann has disappeared among the rows. I know she is there, because her car is parked by the garage. But she is submerged in the green, until I see that familiar hat brim rise above the leaves.

The world is a different place, on my knees. It is smaller, and slower. My vista is all of twelve inches as I look hard to distinguish a mature snap pea from the curled new leaves on the vine. I do not, I cannot, look down the hundred-foot row, or I will lose sight of these peas ready to be picked.

On my knees, I meet the tiny green grasshopper who by fall will be brown and angular, as long as my index finger. The small Trichogama wasp tickles the hair on my wrist as she flies to the nearest flower. I am happy to greet her because she will devour the aphids and deposit her eggs in leaf-eating caterpillars. I disturb the worm who works hard to break through our clay soil. I see the grass blades shiver, and glimpse the brown and black bull snake slithering away, as silently as he can.

As I inch my way down the row, my knees are saved by the bright blue foam pad that I kneel on, but my left hand gets raw as I lean on it each time to shift my weight and push along. I can feel how the earth changes every ten feet or so. It is soft and loamy here, and my hand is grateful, but the dirt is like rocks there, so I must put on my left-hand glove to save my skin. In this section are the thick, dried stems of the winter squash, in that section smooth river rocks and small chunks of quartz left behind centuries ago when the last glacier left.

On my knees, I begin to see with my hands. I can feel the plump stems of the foxtail grass and, without looking, distinguish it from the rounder, smoother stem of the dill. Lamb's-quarters are hard and square, pigweed fat. I've pulled so many of these stems, that all I need do is reach into a clump and feel their texture and shape.

As I pick the shelling peas, I lift the vines that have fallen over, and the coolness under the vines gives me relief from the burning sun. The soil is still wet under the cabbage leaves, while the rows open to the sun are dry and dusty. I feel the wind shift, and the air becomes cooler. Rain coming on. When I'm on my knees, the grass bends over my shoulders and protects me from the rain.

In 1764, Hargreaves invented the Spinning Jenny, making it possible to spin thread twenty-four times faster than with the conventional spinning wheel. In 1890, Frederick W. Taylor systematized work and production at Bethlehem Steel. Taylor was working at "The Steel," as it is called in the Lehigh Valley, as a consultant. He studied the work of the men who shoveled coal into the numerous blast furnaces. He redesigned the yard to save steps, made use of shovels of different sizes and shapes for different materials, and planned the work in advance. These changes made it possible for 140 men to do the work of 400 to 600 men. Taylor saved The Steel an average of $78,000 a year.

These two events—the invention of a machine and the invention of industrial management—are generally considered milestones of industrial progress. They are lauded in a culture that values above all else efficiency, speed, and profit.

It is a different world here, down on my knees. If I went any faster, I would miss the worm, and overlook the pea. I would lose touch with the ground I walk on, or crawl on, as the case may be.

—Olivia Frey, Valley Creek Community Farm, Northfield, Minnesota. Used by permission.

The more clearly people understand what they are getting into when they join a CSA, the greater the likelihood that they will not be disappointed. In 1996, Deborah J. Kane and Luanne Lohr did a study on "Maximizing Shareholder Retention in Southeastern CSAs: A Step toward Long-Term Stability," which sheds some useful light on member expectations. The authors did telephone interviews with new sharers in seven different CSAs before the beginning of the season and again at the end. The interviewers found that "the tone of the fall conversations was considerably more subdued than the spring conversations." The contrast between the spring and fall remarks highlights the kind of misunderstandings that can occur:

Spring:

"I'm looking forward to working with new vegetables that I've never eaten before."

"I'm assuming that these will be pretty much the only vegetables I buy."

Fall:

"When I said I wanted variety, I really meant within the things I was used to eating."

"I never got a wide enough variety to really keep me from having to go to the grocery store."

The percentage of those interviewed who answered that they would pay more for the shares dropped from 66 percent in the spring to 39 percent in the fall, when 23 percent said they would only pay less. Yet in response to the Kane and Lohr questionnaire, of 196 respondents, 121 said they would rejoin, for an overall retention rate among the seven CSAs of 63 percent. Individual CSAs did better, ranging up to 72.5 percent for New Town Farms, which members also rated highly for quality, quantity, freshness, and appropriate pricing of produce. Kane and Lohr conclude with a list of recommendations for increasing CSA retention rates:

- offer rebates to members who refer new members
- get sharers to do the recruiting
- ask sharers to have friends or relatives pick up their shares when they are out of town
- use T-shirts or tote bags with the CSA logo for free advertising
- target local environmental groups, congregations, or businesses in recruiting
- collaborate with other farms
- offer U-pick or some form of choice in the produce
- emphasize the basic crops, with exotics as the occasional treat
- extend the season
- offer value-added products
- offer trial periods
- set up a buddy system pairing experienced members with new ones
- communicate through meetings, resource booklets, crop lists, calendars, newsletters, cookbooks, and marketing materials
- ask members what they want through voting sheets, surveys, suggestion boards, and boxes

I would add to their list:

- offer members choices of as many different levels and ways to get involved with the CSA as possible: from once-a-year farm visits to core group membership to regular farm work
- try to involve members in setting long-term goals for the farm
- include a good question-and-answer sheet about CSA in general and yours in particular
- offer bulk orders of produce from the farm to freeze, can, or otherwise preserve with directions on how to do it
- offer educational programs, farm tours, or on-farm activities for member children
- welcome members to the farm for strolls, camping, skiing, and the like.

CLEMENS KALISCHER

Farmer Elizabeth Smith dances with CSA sharers at Caretaker Farm Harvest Festival.

After seven years' experience chatting with members of the GVOCSA while overseeing distribution, Marianne Simmons has developed a theory about why some people just don't fit. She has observed a crucial distinction between what she calls deductive and inductive cooks. The deductive cook starts a meal by planning—looking up recipes, checking the kitchen for stocks on hand—and then goes out and buys ingredients. Unless these deductive cooks are willing to do their planning on the night they get their vegetables, Marianne explains, the uncertainties of CSA supply will make them very unhappy. They might last a season or two as members or even drop out after the first month. The inductive cook, by contrast, "flings open the refrigerator one hour before dinner, rejoices at the abundance, puts three-quarters of the meal on the table raw, and is happy that this is possible without stepping a foot in a supermarket." Marianne describes herself and her CSA friends as cooks of this kind, and says they are content to let someone else give them a few surprises in their meals. Recruiting new members by word of mouth seems to work best, Marianne believes, because friends and colleagues have a chance to sort themselves out in advance, since they know what to expect. No advertising can be as effective. People who eat out often have trouble remaining in a CSA because they don't do enough cooking.

Marianne believes promoting return membership starts with quality produce and evolves from there. The CSA made it easier for her to get the organic vegetables she wanted to feed her two sons. The local food co-op did not provide a reliable supply at predictable hours. Buying what is produced locally has become increasingly important to her. "If someone down the street made shoes, I would wear them whatever they looked like," she explains. She has observed that other members, attracted initially by the food, develop an affinity for the farm and the farmers. The chance to work at the farm is, in her words, "a gift to the family." Creating an atmosphere that is friendly and tolerant is very important, she points out; a deeper understanding of what sustainability means takes time for people to grasp.

Although we have found word of mouth to be the most effective way of recruiting new members, at least once a year a few of the members and I make an appearance on a radio or TV talk show. The local public radio station has several sympathetic hosts who are willing to have us. Listening members can help by calling in with questions. During one of these shows, I recognized the thinly disguised voice of core member Pat Mannix with a leading question that gave me the chance to plug our next orientation meeting. Lila Bluestone, another core member, volunteered to appear with me on a TV talk show. Before the interviews began, Lila was terribly nervous, repeating over and over the phrases we had agreed to emphasize. But in front of the camera, she blossomed. When the interviewer opened the show with a description of the CSA as a program for "needy" families, Lila corrected him ever so gently, saying, "Yes, the CSA provides the connection with the Earth and the clean, nutritious food of which so many of us in Rochester, of all income levels, are in desperate need." The wags at the next core meeting dubbed this the Liz and Lila Show.

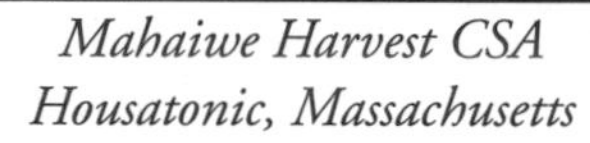

Mahaiwe Harvest CSA
Housatonic, Massachusetts

The organization of the GVOCSA core group has been the main factor ensuring flexibility and endurance. Shared out among fifteen to twenty people, both the physical and the emotional work of maintaining the group do not become burdensome. (See chapter 6 for more details.) The stories of other CSAs are strikingly different. *Farms of Tomorrow Revisited,* by Trauger Groh and Steven McFadden, has some important lessons for understanding what enables CSAs to last. The authors went back to the CSAs profiled in the earlier edition for updates after seven more years. I interviewed some of these farmers or their sharers after yet another year.

COMMUNITY FARMS

Since the publication of *Farms of Tomorrow Revisited*, the Temple-Wilton Community Farm, Trauger Groh's home CSA and one of the two oldest CSAs in the United States, has cut back to thirty-three member families. Development pressure and the lack of long-term tenure of high-quality land has led Trauger, Lincoln Geiger, and Anthony Graham to dissolve their joint farming association. Nevertheless, deeply committed members hope to find a way to enable Lincoln and Geiger to find adequate land for the future. With help from Patti Stanko and Prentice Grassi, Trauger continues to farm the original 22 acres under the name "The Community Farm." Trauger considers all the members as partners in the farm, though not all of them participate in the farm work. (For the story of Buschberghof, the farm in Fuhlenhagen, Germany, where Trauger farmed for fifteen years before coming to the United States and the model for Temple-Wilton, see *Short Circuit*, by Richard Douthwaite).

MAHAIWE HARVEST CSA, which began at Indian Line Farm in 1985, thrives today because it has gone successfully through a transition from the cooperative effort of a group of part-time farmers to one full-time farmer, and from more active member participation to less. Charlotte Zanecchia, a member through the entire thirteen-year history, says that "the key thing is to have a farmer who is happy and keeping his customers happy." Initially, the CSA leased its land from Robyn Van En. By the time the lease expired, Charlotte and some other core members were convinced that the CSA needed to have its own land, unencumbered by individual private ownership. They reorganized themselves into the board of Mahaiwe Harvest Land Trust, gained nonprofit status as a 501(c)(3) and, with the help of American Farmland Trust, negotiated the purchase of a fifty-five-acre farm with a farmhouse. The asking price on the land was $325,000. The new land trust was able to purchase the land at half-price when the Commonwealth of Massachusetts agreed to pay $165,000 for the development rights under the Agricultural Preservation Reserve (APR) program. Before leaving for Switzerland, Charlotte spearheaded a drive to raise the last $75,000 to pay off the mortgage.

For a few years after the move to the new farm, several farmers continued to grow the food for the project. In 1993, David Inglis took the job as head farmer and, as the other growers left, gradually took over the entire farm. At present, David and his family do all of the bookkeeping and share selling, manage a buying club with Northeast Coops for the members, and run whatever events occur at the farm, while he and his apprentices produce all the food for the shares. David has focused on building the soil and improving the quality of the produce, while offering members more choices, but dropping chicken and egg shares until he can do them properly. As a result, membership has grown and attrition from year to year has decreased. The core group has retired, and the board of Mahaiwe Harvest Land Trust has turned its attention to new projects while entrusting the farm to David. He honors the role of the Indian Line core during the early years for creating the model of CSA and for ensuring the continuity of the group through the move. The present form of Mahaiwe Harvest CSA, however, suits his personality and satisfies the members. David is very emphatic on this question:

> It does not really matter to the *integrity* of your CSA whether you have an active core group, no core group, more festivals than you can shake a fist at, or no festivals. The essential point is that it is by mutual arrangement, that you as the farmer and the members of the farm have come to by a process that is shared. . . . What do you do if you don't want to play that [organizer's] role? Maybe you want to grow the best vegetables that you know how to grow, for people you love to grow them for. I'm here to tell you that that's *okay*.

Charlotte recalls "heart-wrenching" moments from the past, but she is pleased with the outcome. In answer to why she has stayed with the CSA so long, she replied:

> I always wanted to have organic produce. The CSA liberated me from having to grow it all myself, though I still garden. I became enslaved to the fundraising for the land. I have had to turn over the land for which I worked so hard to raise money to the farmer. Letting David have his space is the best gift I could give him. But when I eat those first spring crops—all the food, really—it is a spiritual experience for me. We are lucky to eat such diverse crops grown locally. And I see that this kind of farming is what more and more young people want to do. So Robyn and I were right, back at the beginning. The reality of it all has been like teaching a baby to crawl and walk. Finally we are in young adulthood. If David Inglis left, he could sell the business. He has a lot of equity. We hold the land, but we would trust his choice of replacement should he ever leave.

BROOKFIELD, THE THIRD EARLY CSA profiled in *Farms of Tomorrow*, suffered through two very difficult years while the marriage of Nikki and Ian Robb, the initial farmers, disintegrated. Having known the Robbs as meticulous farmers, I realized there was serious trouble when I visited the farm on a NOFA summer tour in 1993 and could barely find some of the crops under the weeds. Under the Robbs' management, as at Temple-Wilton, members were asked to support the entire farm without regard to the precise amount of produce they received, at the cost of $450 per adult. With Dan Kaplan as farmer, beginning in 1995, Brookfield has revised this community farm model. Members purchase a share, more like other CSAs. The Board of Directors of the Biodynamic Farmland Conservation Trust, to which the former farm owners, the Fortiers, deeded the farm, has also introduced a business incentive into the farmer's pay scale. The board pays the farmer a base salary of $15,000, with incremental increases based on how many members he attracts and retains, and how well he manages the business.

Dan manages Brookfield well. With the blessing of his Board of Directors, he has lowered the share price to $300, increased membership from 150 to 350, and introduced the "Mix-and-Match" system from the Food Bank Farm, providing maximum choice of produce for the members who pick up at the farm. When I visited again during the summer of 1997, tractor cultivation had eliminated the weeds. Dan goes out of his way to make the members feel welcome to drop by for a walk or bring their children to watch a calf being born. Brookfield holds regular festivals, but at the Northeast CSA conference, Dan said something very funny about them. He said that members like to *know* that festivals are going on, even though they don't *attend* them much! As Dan sums up the past eleven years:

> Things have changed. . . . We've lost the romantic illusion of community, but gained a real community. Brookfield Farm is not seen anymore as something special and "apart from the real world," but rather just as part of the community, the way the local churches, or schools are part of the community. We are here. We are not the focus of the community, but we are a part of it. (*Farms of Tomorrow Revisited,* p. 143)

The other CSAs profiled in the original *Farms of Tomorrow* have suffered fewer upheavals and have been able to concentrate their energies on steady improvements. Kimberton, Roxbury, and Caretaker farms all pay respectable wages to their farmers with health insurance and even a pension fund. Through U-pick or pre-order systems, they have found ways to introduce more choices for their members. Roxbury sponsors a buying coop and participates in CRAFT, an apprenticeship program. (See chapters 7 and 19 on for more details.) Kerry and Barbara Sullivan at Kimberton CSA made the difficult decision to raise their share price so that they would earn enough to live on. Predictably, they lost some members, but Barbara Sullivan says:

> We ended up with the serious, committed people who really appreciated the garden for what it is. We lost the people who were just looking for cheap vegetables. The whole community has had a really different feeling since then. It was the best move we ever made. We could not have survived until we made that decision. You can start out the way we did, but sooner or later, you've gotta have enough people who are really committed to making the CSA work. It's a real community garden now. We could go, and the Kimberton CSA Garden would still go on. This is a real people garden. (*Farms of Tomorrow Revisited,* p. 155)

SUBSCRIPTION FARMS

CSAs that follow the subscription, rather than the community farm model, have also found ways to solidify their existence. (Subscription CSAs supply members with a box of produce each week and most ask for only payment—not a work requirement—in return.) In its sixth year in 1998, Full Belly Farm in Guinda, California, has reached maximum capacity at 600 shares, and has a waiting list of 150. With year-round production, they offer an enviable selection of vegetables, fruit, and nuts, as well as separate flower and lamb shares. Though most of their recruiting is by word of mouth, the turnover rate is high. The farm is working towards getting all members to make more of a commitment by paying quarterly, instead of monthly. Full Belly welcomes visits by members, but few come. To keep in touch, the farmers do occasional visits to the porches where members pick up shares, taking along a lamb or other attraction to ease conversation. As farmer Dru Rivers says, "The vegetables speak for themselves." A few members though, say the weekly newsletter is even more eloquent than the vegetables.

The more information a CSA can share with members, the better off it will be. Face-to-face interactions are, of course, the most effective. Farmers should take every convenient opportunity to meet with members, hear their ideas, suggestions, and complaints, and in turn tell them about the realities of the farm.

Seven to eight hundred CSA shares represent just a quarter of Be Wise Ranch's production. Bill Brammer added a forty-member CSA to his wholesaling as a community service in 1993. The growth to eight hundred has been almost entirely by word of mouth. People regularly call the farm to say they have eaten Be Wise food at a friend's house and that "it's the greatest they ever tasted." Bill invites them to sign up for a four-week trial period, after which 20 to 30 percent say it is too much food. They switch to buying organic produce at the local health food store—which is okay with Bill, since he supplies most of the produce there too. The only regular member involvement is through hosting drop-off points: Bill cuts the host's price in half for a ten-share site and gives a free share for twenty. The host makes reminder calls to those who forget to pick up. Instead of a frequent newsletter, Bill gives new members an elaborate packet of information when they join, posts announcements at pickup sites, and provides recipes on the farm Web site. After many years as president of the California Certified Organic Farmers (CCOF), Bill is happy to make decisions without the benefit of group process. Nevertheless, he finds his relationship with members "very satisfying." He says it is "fun and a challenge to grow this much food," and he enjoys the many letters of appreciation and calls telling him the kids ate all the Sugar Snap peas before they even got the share home.

Inspired by the book *The E-Myth Revisited: Why Most Small Businesses Don't Work and What to Do About It* by Michael E. Gerber, the farmers at Angelic Organics in Illinois made a careful study of each job on the farm in order to create a clear management structure providing each worker a thorough understanding of his or her responsibilities. Though not writing about farms, Gerber is astute in analyzing how small businesses can completely eat up the people who start them. His book offers a systems approach to making your business franchisable: "The Entrepreneurial Perspective . . . views the business as a network of seamlessly integrated components, each contributing to some larger pattern that comes together in such a way as to produce a specifically planned result, a systematic way of doing business" (p. 72). Spreading franchised Old MacDonald's CSA farms around the country may not be our mission, but Gerber does have some very useful pointers on detaching the function of a job from the personality performing it. (Constructing a drive-through section to our new packing shed, we started to fantasize about the possibilities of a "drive-in" CSA.)

Angelic Organics' goal has been to free up the staff for other creative work besides the farming. They were not able to do this fast enough to prevent one of the initial farmers, Kimberely Rector, from having to go elsewhere to pursue her calling as a graphic artist. But they are clearly making headway: share numbers have grown to over eight hundred; they have enlisted member help in purchasing more land; an increasingly active core group is producing a Food Book, arranging for low-income shares, and helping with member recruitment and farm financing; and they continue to publish an outstanding newsletter. The CSA has generated enough revenue to begin making up for two decades of undercapitalization. That, of course, is one of the major obstacles preventing farmer John Peterson from doing more of his wonderful essay writing. He currently is busy building infrastructure, creating systems, and upgrading equipment. The Angelic Organics core group has also launched a quarterly newsletter of its own, prophetically entitled "A Farm Forever." On page 1 of volume 1, number 1, Farmer John lists some of the most burning questions confronting CSA sustainability:

> How do we truly build community? How do we take care of our shareholders beyond raising wonderful vegetables and delivering them on time? How can the financially disadvantaged have access to our vegetables? What needs to be done to put the farm on solid financial footing? How does the farm secure an adequate land base? How does Angelic Organics become a CSA that can survive the eventual or sudden loss of its managers and/or its head farmer? How can we share the farm with our shareholders and the general public through experiential education and other programs?

When we can answer all of those questions, we will know the future of CSAs.

Jered Lawson, a close observer of and technical adviser to CSAs, has expressed the concern that the larger subscription projects will crowd out smaller ones, while not providing a meaningful connection between the sharers and the farms. Watching CSA membership on a few farms grow into the hundreds, Jered feared that CSA might fall into the clutches of agribusiness as usual, with the farmer behind a desk, labor reduced to the status of "input," and expansion and profit as the main goals. The evolution over just a few years of Full Belly, Be Wise, and Angelic Organics suggests that size and the subscription model do not necessarily destroy the possibility of meaningful farmer-member partnerships. With time and creativity, the larger farms may yet figure out how to provide their membership with, in Jered's words, "the opportunity to keep moving toward greater conscious support of their farm's, and society's, general economic, social, ecological, and cultural health" (*Farms of Tomorrow Revisited*, p. 86).

COMMUNICATIONS

John Peterson's writings and the Angelic Organics newsletter offer another important clue as to how to maintain a vital CSA: good communications. Steve McFadden, one of the authors of *Farms of Tomorrow,* goes so far as to say that "those [CSAs] are most successful that communicate most clearly." Whatever medium you use, the more information a CSA can share with members, the better off it will be. Face-to-face interactions are, of course, the most effective. Farmers should take every convenient opportunity to meet with members, hear their ideas, suggestions, and complaints, and in turn tell them about the realities of the farm. Once you move into signs, flyers, e-mail, Web sites, and newsletters, you venture onto somewhat less predictable ground. Participants in a workshop on communications at the Northeast CSA conference, after a discussion of effective signage, concluded that there is no overestimating people's capacity to *not* read signs. Where you must use signs, fewer is better, and the lettering must be large and clear. Scott Chaskey described the system at Quail Hill, where all 150 members pick for themselves: right out on the beds, they post on wooden stakes computer-generated signs with large, emphatic lettering stating "Pick Only Outer Leaves!" "6 beets," "12 tomatoes."

Cheap Food and Flat Tires

Erik [an intern at Angelic Organics] gets up at 2:00 in the morning twice a week to get the veggies to you on time. He checks shareholder names against a flow chart as he loads a hundred boxes ($1^1/_2$ tons) of produce and numerous buckets of bouquets from the cooler into our blue truck by flashlight. Coffee mug and cellular phone beside him, he guns out of the farmyard at 3:45 A.M.

The last time Erik headed the blue truck towards Chicago, the right front tire shredded. It was 4:30 A.M. Erik gulped coffee, then unloaded half your boxes to get at the spare. He bounced up and down on the breaker bar to remove the stubborn lug bolts. He changed the tire. Although heady with exhilaration from wresting those lugs to the ground, Erik did not forget to reload your boxes before again heading toward the Big City.

We've had a lot of flat tires this year. It's because the roofs are going. The winds zing shingles onto the yard and into the flowers. I find roofing nails in the driveways. Yesterday I found one on the kitchen floor. Someone must have tracked it in.

The last roofing we did here was in '74. We did all ten buildings. Then, I farmed big until the money ran out. In the mid-eighties the roofs started to leak—slow leaks at first. It took a blast of a storm and gusty winds for a meager drip to find its way inside. Now a mere drizzle can puddle the floors. When I pick up the phone after a rain, the receiver is sometimes cradled in a little pool of water.

Once the countryside was boisterous. Barns were red and straight. Livestock strutted in the farmyards. Children played with ducks. Now—take a drive from Chicago to our farm. Look for farmers walking in their fields: They are not there. Count the children playing in farmyards; where are the children? Look at the farmsteads. The paint goes. The roofs go. The ridgepoles go. The countryside is desolate, propped up by a mock luster of chemicalized corn and soybeans, and occasional gentrified hobby acreages.

I can't isolate the causes of this dilapidation of our countryside. I know about mechanization and computerized farming and the millions of tons of chemicals that seem to have stripped the people from the land, but what causes what, anyway? What caused the willingness to have the barrenness occur this way?

People work hard on this farm. There's barely enough money to pay them. There's barely enough money for the myriad other immediate expenses that accrue to this small but complex operation—the expenses that somehow get the operation into the next month. I know other farms have seen their roofs going and couldn't bear it and sold out. Farming our way into poverty amongst a hail of rotting shingles and roofing nails while affluent consumers stalk food on the cheap yet demand the most expensive "health care" is an insult. Getting a flat tire in the middle of the night because our roofs are blowing away just might have some relationship to the cheap food mindset that pervades this country.

Community Supported Agriculture is a step toward restoring this farm. It offers a deeper relationship than can be achieved through money spent at the organic counter for produce from "somewhere in California." But that California produce influences the price we charge for our memberships. We are often asked how our prices compare to store prices. Large, heavily capitalized organic operations in a remote temperature climate set the price standard for Midwestern organic farms. The true cost of sustainable farming—a farming operation that really takes care of itself, its machines, its buildings, its land, its workers, its future—is far from the cost of our shares.

—John Peterson,
Angelic Organics,
Caledonia, Illinois.
Used by permission.

For all CSAs, but especially in those where members have few personal contacts with farmers, a regular newsletter is an essential part of communications. The newsletter does not have to be fancy. West Haven Farm posts weekly news to most of its members by e-mail. Many CSAs dash off weekly one-pagers, some of them handwritten. Willow Pond has a regular format with a list of the season's vegetables along the top. Farmer Jill checks off what's in the share that week, does a quick drawing of one vegetable with a note about its nutritional value, fills in a few lines on "Upcoming Events," and jots down a recipe in the box underneath. A tear-off section at the bottom invites member feedback.

Hidden Valley Organic Farm has a two-pager, front and back. Like Willow Pond's, it includes a list of all the vegetables with check marks by the selections of the week. Most of page one reports on the state of the crops, the weather, things to come, and special requests from the farmers. Recipes cover the back page. Moore Ranch CSA puts out "Harvest Notes," a more polished two-pager desktop-published on a computer, with weekly share contents, short notes about the crops or the state of the farm, an essay on Biodynamic growing, cooking or preserving vegetables, and a column that focuses on a particular vegetable family, such as brassicas or alliums, with some recipes. The back page carries reprints from the Biodynamic Association Newsletter on subjects such as "The Role of Animals," or "The Mystery of Form in Nature."

Good Humus Farm
Rochester, Washington

Stephen and Gloria Decater publish a more elaborate, but less frequent newsletter for Live Power Community Farm, illustrated with photos of farm animals, children, and festivals. Cumulatively, the Decater's articles provide a narrative of their lives on the farm through the seasons.

I haven't read all CSA newsletters, so I may be judging unfairly, but surely the Pulitzer Prize must go to Angelic Organics. Edited by farm office manager Bob Bower, this newsletter absolutely sparkles with writing talent, (both Farmer John's and others'), energy, good humor, and information about food.

Editing and mailing a regular newsletter is an excellent way for members to contribute to their CSA. If I am any example, it is not hard to find non-farmer members with better computer skills. Since its second year, the GVOCSA has had a newsletter that appears five times: four issues during the growing season and one in the winter. We had the good fortune to have Pat Mannix as editor for five years. (See her essay "Faith Communities on the Land" on p. 206.) I have had the pleasure of contributing "Notes from the Farm," with some help from David—a continuing series of reflections on the crops, the weather, the people, and the critters at Rose Valley Farm. The editor and other members contribute messages from the core group, articles on topics such as home schooling, how one person consumes an entire share, or the global significance of buying local, as well as recipes and notices of events. A sixth annual mailing to members is a fat packet of information that includes the membership list, a list of core members with their responsibilities and phone numbers, the work schedule, a map to the farm, maps of the farm and distribution sites, a page of suggestions on how to be ready for farm work, and the bulk order price list.

The GVOCSA also encourages members to purchase the *FoodBook for a Sustainable Harvest,* which David and I wrote during the winter of 1993 as a guide to the selection of vegetables, herbs, and small fruit we provided in the shares. Presented by season, each entry includes a little history of the crop, nutritional information, how to store for the short term and long term, anecdotes from growing at Rose Valley Farm, and recipes. Proceeds from sales of the *FoodBook* go to the GVOCSA scholarship fund for low-income members. I added one hundred more recipes from GVOCSA members and other CSAs in 1996. (See the CSA Resources section.)

The CSA newsletter is blossoming into a new literary form, unveiling previously undiscovered talents. One member of Harmony Valley CSA commented: "I can't decide if the newsletter or the produce is the best thing you do—but since we can't eat the newsletter, I guess it comes in second. . . ." Annie Main of Good Humus Farm says that after twenty-two years of farming, the necessity of writing weekly essays has changed the way she experiences her life and work: "I live it looking for things to write in the newsletter. It has heightened my awareness of where I'm at." In anticipation of a future collection of CSA essays, I have chosen a few of the best examples to whet the reader's appetite (see the boxes sprinkled throughout this and other chapters).

In addition to a newsletter, Harmony Valley gives its CSA members a calendar packed full of useful information. Noted on the appropriate dates are farm events open to members, the farm's work schedule, including planting and harvesting days, payment reminders, holidays, the farm family vacation plans, and, at the end of the year, reminders to sign up for the next year. Most of the illustrations are lovely nature scenes or vegetables, but both May and July have diagrams on how to flatten the box used to deliver members' vegetables. The cover bears a map to the farm and the words, "You are always welcome!" (A cautionary note on open invitations: "always" means seven days a week, twenty-four hours a day. You can be very welcoming to members, while restricting visiting hours to the times convenient for the farm.)

At the pinnacle of up-to-date communications, CSAs with excess computer energy are creating Web sites. On the Internet, a central CSA site, sponsored by prairienet.org, provides a definition of CSA and a series of links to CSAs with sites of their own. Angelic Organics has one, of course, where the curious can view the map of pickup sites around the Windy City. They are adding recipes, a chat room, a bulletin board for members to exchange information, and farm photos. Michael Axelrod, a Web wizard in the GVOCSA, set up a site for us in 1996, where anyone on the globe who is appropriately wired can read about how to get our vegetables and access the archive of my "Notes from the Farm." Catalpa Ridge Farm, in Newfoundland, New Jersey, has an extensive site replete with photos of farmer Richard Sisti building a new greenhouse, a page on the farm's training program for handicapped adults in the Easter Seals Highlands Workshop, and a dazzling array of the twenty red and blue ribbons the farm vegetables won at the Sussex Farm and Horse Show. (If you detect a slight note of sarcasm in this paragraph, you will be right. I confess that the Luddite in me is not yet convinced that we need the World Wide Web to promote our local vegetables.)

Yet another dimension of CSAs is the role they can play in agricultural research. Harmony Valley offers its members a "unique opportunity to support some cutting edge research" in progress on the farm. Farmer Richard de Wilde has been working on demonstrations of the use of hedgerows for habitat for beneficial insects, and of compost for controlling plant diseases. Christine Gruhn, a member of the GVOCSA and a mycologist, is studying the mycorrhizae—the symbiotic association between fungi and plant roots—that help nourish our crops. As part of a course in the scientist as citizen, she had her students grow lettuce and collards at the farm to donate to the local food bank. The Practical Farmers of Iowa, in cooperation with Iowa State University, have research projects on several CSA farms. Both the Sustainable Agriculture Research and Education Program (SARE) and the Organic Farm Research Foundation offer research grants to farmers for projects of this kind. (See CSA Resources for addresses.)

COMMUNITY BUILDING

I really just wanted to write a few words about farm festivals. But I have heard so many farmers whose on-farm events have been poorly attended express disappointment about community-building that I find myself compelled to address the much broader subject of the meaning of community. The number of people dancing around a maypole, roasting marinated vegetables over an open fire, feasting together on fresh strawberries, or taking turns on the crank of a cider press—is that the true indicator of community?

Certainly, some CSA farms hold fabulous festivals. Two thousand people attend the Hoes Down Harvest Festival at Full Belly Farm, run with the volunteer help of CSA members. Out of 400 families, 125 come to strawberry picking day and 80 to 100 to garlic braiding day at Harmony Valley Farm. In their first year, Seeking Common Ground CSA could not promise large amounts of produce, so they supplemented the shares with conviviality and celebrations. They held weekly courses and found a theme for a festival every single month. For years, Rose Valley Farm held an annual potluck breakfast, maypole dance, and wildflower walk, starting at 7:30 A.M. on the Sunday closest to May 1. No matter how cold and rainy the weather, we always attracted enough neighbors and CSA members to hold every ribbon on the pole. David Inglis, who has set in abeyance some of the festivals once held at Mahaiwe Harvest CSA until members come forth to arrange them, still enjoys the "burning of the hat" ceremony at the end-of-the-year party he organizes himself. Children love to see him burn his hat.

Distribution time, when members converge on a pickup site, provides the occasion for many CSAs to foster a sense of community. The GVOCSA holds "pickup parties" once a month at distribution where members are lured with name tags and snacks to linger and socialize. The Magic Beanstalk does its distribution in a church basement where other activities are also going on, such as the sale of other farm products, games for children, refreshments, and information tables, and where groups of chairs are arranged to invite members to sit down and interact. Farmers who make deliveries to CSA pickup points distant from their farms often stay for the distribution time to socialize with members.

Many CSA farmers have also found creative ways to share the natural beauty of their farms and agricultural knowledge with members and the general public. Several Biodynamic CSAs have close relationships with Waldorf schools and regularly host programs for the children. Winter Green Community Farm in Oregon and Clark Farm in Connecticut give tours and classes for local public-school children. Caretaker Farm in Massachusetts, Willow Pond Farm in Maine, and Willow Bend Farm in Minnesota have special gardens by and for the children of members. Happy Heart Farm in Colorado welcomes student visitors from around the world. One day a week, three handicapped adults help Richard Sisti at Catalpa Ridge with transplanting, seeding, weeding, and harvesting. Over the years, dozens of classes, from kindergarten to graduate students, have toured and helped with work at Silver Creek and Rose Valley farms.

> **In surveys of CSA members in all regions of the country, the desire for "community" ranks very low among the reasons for joining. Most people are not looking for more busy-ness in their lives. . . . Perhaps we need to think about community differently, as a sharing of values and a commitment to act upon those values, even in very modest ways.**

More Than Good Eats

Some of you, perhaps, did not fully grasp what you were getting into when you signed up to receive a box of vegetables each week for twenty weeks. You did not realize that you were signing on to have your week's meal plan monopolized by fresh vegetables demanding to be cooked and eaten before they pass their prime. That you wouldn't have as much room for dessert after a big dish of Pasta Primavera made with your abundance of fresh organic veggies. That a grocery store tomato would never satisfy you once you had tasted our farm grown, freshly picked, vine-ripened gems.

But did you stop to think about the further implications of belonging to a CSA? In supporting Angelic Organics, you are not just buying a commodity. You are supporting a dwindling way of life, the education of future farmers who learn here, a sense, even, of mission—which regards farmland as worth something more than its real estate value.

Sure, the money you pay goes towards the production costs of organic vegetable raising. Fuel, labor, seed, soil amendments, and plant protections such as row covers are provided for by your checks. But your money also enables relationships. It backs Farmer John's relationship with this, the land he grew up on. It allows him to continue labor negotiations with his machinery, coaxing it to help him out for one more year. It enables his relationships with those who work for him and those who buy from him.

You might come out of this year with new favorites, and familiarity with obscure vegetables with which you can impress your friends. ("Yes, the Rutabaga Bisque I made last week was absolutely delightful, darling.") You may have much stronger opinions about what constitutes "yucky" and "yummy" in the vegetable kingdom. You may pick up a few new recipes, or make up some of your own. But you can take your membership further than that if you wish.

Community is a word with potential. Community can simply mean the people who live in a selected geographical area. But it can also mean those with whom one has a more meaningful relationship than simple geographical proximity. You may find yourself swapping recipes with other shareholders when you pick up your box. You might wish to come out and volunteer for a day or more on the farm; get a taste of country/break from the big city. You might wish to join our new Core Group, which was founded this year to take over the farm's non-production tasks such as marketing, special events, and surplus vegetable distribution to the needy. Engaging yourself in one of these ventures, you are at risk of discovering within your heart the feeling of community.

Community in this sense is not a commodity that is purchased; it is built through actions. We at Angelic Organics propose to bring fresh healthy produce into your life. Nothing to sneeze at in and of itself. We also give you the option of taking your membership further; to make your involvement with Angelic Organics encompass not only what you pull out of your vegetable box, but what you put into your CSA.

—Kristen the Cook, Angelic Organics Farm News, Caledonia, Illinois, July 1997. Used by permission.

Nevertheless, in surveys of CSA members in all regions of the country, the desire for "community" ranks very low among the reasons for joining. (Cohn, Cone and Kakaliouras, Kane and Lohr, Kelvin, Kolodinsky and Pelch.) Most people are not looking for more busy-ness in their lives. Squeezing in the time to pick up the food amidst all their other obligations is hard enough. As Cynthia Cone and Ann Kakaliouras put it in their article, "The Quest for Purity: Stewardship of the Land and Nostalgia for Sociability": " 'Community' for many [CSAers], if not most, seems to be an expression of longing, a nostalgia for imagined linked relationships of our rural past—a kind of community that is difficult for CSA members to realize given the demands and constraints of their lives" (*CSA Farm Network*, vol. II, p. 29).

Perhaps we need to think about community differently, as a sharing of values and a commitment to act upon those values even in very modest ways. In his master's thesis, Tim Laird quotes a farmer who comes closer to this understanding of community as "the people's involvement, helping at the farm with promotion, making friends, creating a connection to the Earth and each other" (Laird, p. 94). Of the farmers Tim surveyed, 63 percent believed they were achieving success at building community in this deeper sense and mentioned that this was the most gratifying aspect of CSA for them. One of the farmers wrote:

> Everyone feels a part of what is being done, what is going on. It is so much more satisfying than money for vegetables. Members have visited the farm in the off-season. They bring their children. They see where the fruit and produce come from and learn the process. We (farmers) meet concerned people who trust us enough to give us their money without contract or guarantees—we care about each other. We are into something mutually upbuilding. (Laird, p. 101)

Jack Kittredge would agree, concluding his essay "Community Supported Agriculture: Rediscovering Community": "To the extent that alternative institutions recenter our lives in small, local groups of people with whom we have mutual obligations, they can rebuild human community" (*Rooted in the Land*, p. 260).

◆

THE DETERMINED WORK of a relatively small group of people keeps each CSA afloat. Local, fresh, tasteful food is the key. But for community supported agriculture to flourish—and not merely survive—many farms and gardens must grow, crosspollinating promiscuously with one another, every good idea lingering in the public sector for each of us to sniff suspiciously and snatch up if it suits us. There is no formula, except a little imagination and a lot of hard work.

— Part IV —

The Food

GROWING THE FOOD

13

This we know: the Earth does not belong to man. Man belongs to the Earth. This we know: all things are connected. Whatever befalls the Earth, befalls the sons of the Earth. Man did not weave the web of life. He is merely a strand in it. Whatever he does to the web, he does to himself.

—ATTRIBUTED TO CHIEF SEATTLE

Eating a CSA share is like having your own garden, but with much less work. For the farmers who produce that share, however, the inverse is true: growing a large variety of crops so as to have an appealing selection and combination every week over a six- to twelve-month season is a challenge. The mix required is similar to growing for a farm stand or a farmers' market booth. Depending upon your previous marketing experience (or lack of it), growing for a CSA may require learning how to produce a much wider range of crops than you have done before. You will need to make careful rotation and crop succession plans to be successful.

Although in theory, a conventional farm could start a CSA, in practice, almost all of the existing CSAs are committed to organic or Biodynamic methods. Several farms, such as Pat and Dan LaPoint's Hill 'n' Dale in Pavilion, New York, have started CSAs while in transition out of chemical use. Two of the seventy-three farms in Tim Laird's survey were not organic. Market studies show that the potential members for CSAs are mainly people who seek "clean" food for themselves or their children. Most of the CSA brochures include descriptions of the farm cropping system. Here are three typical ones, gleaned from the hundreds of brochures I have read:

COMMON GROUND FARM in Olympia, Washington: "We farm organically, though we are not 'certified organic.' We grow nutritious vegetables by attending to the health of the soil and plants. We use composted manures and cover crops to maintain fertility, and minimal tillage with light machinery to avoid compaction and erosion. Damage from pests or disease is minimized through our crop selection and rotations, and the judicious use of row covers. Increasing local food self-reliance is part of our work, and toward that end we trial vegetable varieties that extend the harvest season both early and late."

GRANGE SUR SEINE in La Broquerie, Manitoba: "We try to run a traditional mixed farm. We have chickens, sheep, goats, pigs, two cows for milk, and horses for working the land. We use solar and wind energy for electricity and water heating. We have been largely self-sufficient by growing a huge garden, and we want to share this with others. We care for the Earth and aim to pass it on to our children in as good or better condition as when we received it."

FIDDLER'S GREEN FARM in Brooks, California: "At Fiddler's Green we are committed to sustainable organic farming practices. Organic means no use of fertilizers, herbicides, nor pesticides that are synthetically produced. Sustainable means we want to continue farming without depleting our natural resources of land and water. Therefore, we rotate our crops, plant cover crops, apply organic compost, introduce beneficial insects, use mechanical and hand cultivation for weed control, select proper irrigation methods, consistently care for our soil, and grow seasonally."

In this chapter, we will not attempt to teach how to grow a wide variety of vegetables and fruit. Use the many good resources available already. For an introduction to organic farming, *The Real Dirt: Farmers Tell about Organic and Low-Input Practices in the Northeast* is a good place to start. Eliot Coleman's books *The New Organic Grower* and *Four-Season Harvest* are essential reading. The brand new book by Vern Grubinger, *Sustainable Vegetable Production on One to One Hundred Acres* is the best single volume for professional vegetable growers, organic or conventional, I have seen anywhere. If you are a beginning farmer, an excellent way to learn the trade is through an apprenticeship or internship with an experienced farmer or better yet, with several farmers. You can also learn a lot by working on a variety of farms, even for short periods. Joining one of the many farmer networks that exist in most states, and attending conferences and farm tours are other good ways to pick up useful information. These days, many conventional growers' conferences often include helpful sessions on biological controls and soil-building techniques. In some parts of the country, Extension agents are knowledgeable about organic farming, a big change from even five years ago.

I learned the hard way by jumping from a large garden right into full-time farming, but I really do not recommend that approach. While it's true that you learn from your own mistakes, you can shorten the list and reduce the pain through hands-on training with someone who really knows the trade/art of farming. I did have the benefit of many trips to France, where I visited small farms and market gardens, and observed some excellent examples of well-run professional operations. By the time I moved to Rose Valley Farm, I had completed the equivalent of an apprenticeship during eight years at Unadilla Farm in Gill, Massachusetts.

At Unadilla, we started with a single crop of leeks in 1981, and then increased to about ten crops on four acres of raised beds. At the same time, we grew a full range of garden crops for our own use. We marketed to restaurants, food co-ops, a farmers' market, and directly to neighbors. We also sold through two grower co-ops. Although we were the only organic growers in the Pioneer Valley Growers Association and took a lot of razzing, I learned a lot about quality, packout, and how the vegetable business works. In 1985, I participated in the creation of Deep Root Organic Truck Farmers.

I had discovered CSA in France in 1977, but my partner in Gill was not interested in trying it. At Rose Valley, David was open to the idea, and his long-standing connection with Alison Clarke and the Politics of Food in Rochester provided a perfect group with which to start. In terms of production and marketing strategy, the CSA meant adding quite a few more crops. David had already established an acre of asparagus and half an acre of blueberries, and was producing garlic, snow peas, assorted greens, broccoli, winter squash, and some root crops. We added shelling peas, herbs, flowers, cucumbers, onions, green beans, and a wider variety of greens.

When arrangements with a neighboring farm to produce tomatoes, summer squash, melons, and potatoes did not work out, we took those on as well. Our friends at Black Walnut Farm were happy to grow these crops, but did not want to take on the organizational work of the CSA.

After nine years, David lost his appetite for juggling that many crops. I still find fun in discovering new ones, especially when I can fit them into established cropping patterns. For example, planting mizuna along with bok choi, Chinese cabbage, and lettuce in the spring, or throwing in a row of senposai next to the kale and collards for the fall involves little additional work. Growing for the farmers' market, we were used to planting greens sequentially every ten days so as to have a steady supply. It took a year or two to figure out that the best planting sequence for carrots was two 300-foot beds every two weeks from mid-April through late July. Supplying a steady flow of broccoli all through the fall is trickier: shifting waves of cold and heat can cause broccoli plants started as much as a month apart to ripen at the same time.

I am still searching for the best sequence for crop rotations. After a lot of reading and seventeen years of experience, I don't think we understand very much yet about how to optimize our rotations. Farmer and gardener observations over centuries confirm the value of rotations for soil fertility, pest and disease reduction, and yield improvement. We have some rough guidelines: do not plant crops of the same family repeatedly in the same field; do not plant crops that share diseases in the same field; alternate deep- and shallow-rooted crops. Another approach is to progress from sod to large-seeded crops or transplants, to smaller-seeded crops that require better weed control. In *The New Organic Grower*, Eliot Coleman suggests a sequence based on his observations, some of which coincide with mine, such as potatoes following corn. But the explanation for why rotations work remains a mystery, perhaps because the factors are too many and complex for the human mind to disentangle. With the help of mycologist Christine Gruhn, we have begun to explore the role of myccorhizae in relation to rotations. It has taken three years to even know what questions to ask. After three years of sampling roots, Christine has data that show that some myccorhizal crops, such as lettuce, benefit from following other mycorrhizal crops, while others, such as garlic, show no difference. We have only scratched the surface of this fascinating puzzle.

As a rule, farms producing for CSAs seem to be able to provide an average of twenty shares per acre. CSA farmers tend to reduce the amount of land needed per share as they become more experienced. Increased mechanization lowers the hand work per share, but also lowers the number of shares per acre. The most intensive farms rely almost entirely on hand labor.

Supplying a CSA gives a grower the opportunity to use varieties that commercial agriculture avoids because they are too delicate to handle, do not ship well, or are not adequately uniform in size and shape. Robyn wanted to include recommended varieties in this book, but I think choice of variety is too region- or even farm-specific for any one list to be very helpful. Among CSA growers there is a lot of interest in heirloom varieties. I overheard two members of my CSA puzzling over what that term means. One asked, "What are heirlooms, anyway?" In a voice of authority, Jeff Mehr replied, "They are members of V.A.R. (Varieties of the American Revolution)." We may need to organize a small army of seed producers in defense of V.A.R. to ensure that open-pollinated and non-genetically engineered seed continues to be available. (see Chapter 22, for more on how growers are sharing seeds, and the essay by C. R. Lawn on the politics of seeds and the seed industry on p. 220–21.)

As a rule, farms producing for CSAs seem to be able to provide an average of twenty shares per acre. Through his repeated surveys, Tim Laird noticed that CSA farms tended to reduce the amount of land needed per share as they became more experienced (Laird, p. 29). Since CSAs range in size from three to over eight hundred members, and the farms cover every level of mechanization, it is hard to generalize. Increased mechanization lowers the hand work per share, but it also lowers the number of shares per acre. The most intensive farms rely almost entirely on hand labor. John Jeavons teaches that with biointensive techniques it is possible to feed one person a vegan diet for a year using only 4,000 square feet. Utilizing a judicious mix of mechanical and hand methods, Sam Smith of Caretaker Farm in Williamstown, Massachusetts, grows enough food for 170 households for ten months of the year on 6½ acres, of which 2 are always under cover crops, an average of forty shares per acre. The Food Bank Farm in Hadley, Massachusetts, supplies five hundred families and gives an equal number of pounds of produce to the Food Bank on 40 acres, for over twenty shares per acre. What these farms have in common is careful planning combined with the ability to be flexible and adjust to unpredictable weather conditions.

While not abandoning the concept of sharing the risk, many CSA farms purchase a few crops from other farmers. The summer Rose Valley was under water, we bought winter squash from our friend Rick

On Successions and "Aesthetic Gluttony"

"Rocket" is another name for arugula. I planted it last Friday, just ahead of rain. On Sunday, it was poking through the dirt. On Monday, it was a row of little dots heading down the field.

The many crops here, with their diversity of rhythms, are a bit like a jazz performance: one crop of potatoes per season; five of arugula; beets—four; winter squash—one; cucumbers—two; carrots—four; mizuna—three; lettuce—five.

We brought a thousand pounds of melons a day out of our sow pasture. The melons are gone now, the field mowed, plowed, disced, seeded to oats and vetch. First planting of cucumbers and zucchini—gone. Onion field—plowed up and seeded, onions in storage. Fields are gingerly tended then pulverized and put to rest, or planted again.

These different rhythms and different shapes of crops offer a dazzling visual feast. Fuzzy rows of carrots streak to the west, flanking scalloped tufts of green and red lettuces. Palm-tree–shaped Brussels sprouts transform a service drive into The Grand Boulevard. Massive cabbage leaves flop in a savoy sprawl on the dirt and gradually hug themselves into a big ball. Blue-green broccoli leaves jut huge and rigid while, deep in the center, a tuft of yellow-green begins its fast journey toward broccoli destiny. Enormous heirloom tomatoes hang voluptuously on avenues of trellising, blushing yellow, pink, purple. Gladiolas spire, then gush red, magenta, white, rose.

To gaze at the whole lush display of textures, form, colors, to notice the daily changes (yes, there are *daily* changes), is a privilege of being a farmer. "Aesthetic gluttony"—AG for short—the team here has dubbed it. Pink morning mists and a long row of east-facing sunflowers ready to greet the day are casually acknowledged: "Really AG."

—John Peterson,
Angelic Organics, Caledonia, Illinois.
Used by permission.

Schmidt. When major crops fail to thrive, such as spring spinach, or winter squash, David Trumble, at Good Earth Farm in New Hampshire, informs his members and buys replacements from other organic farms. If less essential vegetables like bok choi or kohlrabi are in short supply, his shares go without. Subscription CSAs regularly fill out their boxes with supplements of crops they do not grow themselves. They pay the going wholesale price for the produce and do not expect to make money on the deal. Each CSA must find the right balance between member satisfaction and the financial value of risk-sharing.

The team behind CSA Works—Linda Hildebrand, Michael Docter, Dan Kaplan, and Scott Reed—have created a planning chart based on the preferences of their combined seven hundred sharers. They have graciously allowed me to reproduce the chart in this book, along with their explanation of how to make the best use of it.

Steve Moore, of Moore Ranch in California and president of the Biodynamic Association, has developed a similar planning system, which he published in the September/October 1997 issue of the association's newsletter. To get a copy, you can call ATTRA, "the national sustainable agriculture information center," at 1-800-346-9140: Steve's article is part of their very helpful CSA information packet.

Share charts from two other farms in Wisconsin and Maine provide a sense of the range of crops and the range of quantities provided by different farms. The chart from Jill Agnew at Willow Pond Farm in Sabattus, Maine, which gives both the projected amounts for three years and the actual amounts for two, is especially interesting. There appear to have been crop failures or serious shortfalls in a few places—kohlrabi and zucchini did not produce for Jill in 1994. And some spectacular bounties—5 pounds of peas planned for 1994 and 40 delivered, 40 pounds of tomatoes planned for 1995, but 103 pounds delivered. Overall, though, an experienced grower like Jill is able to predict production levels with a remarkable degree of accuracy, while caring for sheep, horses, chickens, turkeys, two children, and a husband.

Once you have calculated what to plant and how much, you need to figure out the timing. Drawing up a careful plan of where to plant each crop based on your rotations is a good winter activity. You also need approximate dates for starting in the greenhouse or direct seeding. Actual timing and placement will depend on real field conditions. I usually feel very successful if I rate 75 percent accurate. Another critical date to note is the outside limit. For example, I learned from hard experience that at Rose Valley, if we planted carrots after the end of July, they would not have time to mature before the end of the growing season. Spinach planted after mid-June is a gamble, unless you have irrigation to prevent it frying in the heat. Brassicas planted after June 20 will not suffer attack by root maggot. These dates, of course, will vary according to your particular site.

Succession planting combined with variety selection makes it possible to spread out the ripening and harvesting of a crop. I usually do two or three plantings of corn at two-week intervals. For each planting, I combine a faster-maturing corn with a slower variety. That way I can usually provide a steady flow of corn over six or even eight weeks. I have also discovered that broccoli started in different ways comes to maturity at different times. I seed the same variety of broccoli in the ground, in speedling flats, and in soil blocks on the same day and set them out at the same time. The field starts mature first, the soil blocks second, and the speedlings last. When weather cooperates, this strategy can give you a steady harvest flow over three weeks from the same planting date. Similar effects are probably possible by varying irrigation. Again, these subtleties may be site-specific.

As a guide to succession planting, Scott Chaskey and Tim Laird at Quail Hill Farm developed a spreadsheet to record crop, variety, greenhouse seeding date, direct seeding date, area or linear feet planted, amount of seed needed, number of plants, number of plants or flats per bed or row, amount harvested per summer or winter share, and comments (see the sample on p. 140). Steve Decater created a similar

CSA CROP PLANNING CHART

Crop	*Per share per year average*	*Yield per row foot*	*Annual goal per 100 shares in pounds*	*Yield per 100 row feet*	*Rows needed @ 100 feet per row*	*Percentage of total acreage*	*No. of weeks harvested*
Arugula	3.74	0.17	374	17	21.99	1.48	20
Basil	0.25	0.05	28	5	5.60	0.38	9
Beans	3.20	0.15	320	15	21.33	4.30	10
Beets	16.28	1.00	1,628	100	16.28	1.64	19
Bok choi	5.00	0.50	500	50	10.00	0.60	14
Broccoli	15.00	0.25	1,500	25	60.00	12.11	11
Broccoli raab	1.25	0.50	125	50	2.50	0.25	10
Brussels sprouts	5.69	1.00	569	100	5.69	1.15	3
Cabbage	19.93	2.00	1,993	200	9.96	2.01	11
Cantaloupe	9.00	4.12	900	412	2.18	0.79	5
Carrots	66.05	1.49	6,605	149	44.35	4.47	18
Cauliflower	9.66	0.84	966	84	11.49	2.32	4
Celeriac	3.00	1.49	300	149	2.01	0.20	2
Celery	2.00	1.50	200	150	1.33	0.13	0
Chard	3.27	1.10	327	110	2.98	0.30	14
Cilantro	1.50	0.40	150	40	3.75	0.25	19
Collards	2.70	1.00	270	100	2.70	0.54	13
Corn	65.00	0.66	6,500	66	99.21	18.02	6
Cress	0.10	0.25	10	25	0.40	0.03	1
Cucumbers	12.00	4.51	1,200	451	2.66	0.97	11
Dill	0.45	0.30	43	30	1.43	0.10	11
Eggplant	10.00	1.50	1,000	150	6.67	1.35	6
Endive	1.20	0.20	120	20	6.00	0.40	10
Fennel	0.96	0.54	96	54	1.77	0.18	3
Han tsoi soi	0.50	0.45	50	45	1.11	0.11	8
Garlic	2.35	0.45	313	45	6.95	0.47	3
Garlic flowers	0.26		26		byproduct		2
Kale	7.08	1.50	708	150	6.18	1.24	16
Kohlrabi	1.91	0.51	191	51	3.76	0.38	5
Leeks	5.00	0.59	500	59	8.42	0.57	4
Lettuce	22.08	0.20	2,208	20	110.41	7.45	20

How to Read the Chart

The first column shows the average amount of each crop consumed per shareholder over the course of the entire season. The second column shows our yield per row foot for each crop. The third column is simply the first column multiplied by one hundred shares, which provides the total amount we need to grow for each one hundred shareholders. On your farm, instead of one hundred, you would multiply by the number of shareholders you have. The fourth column, "Yield per 100 row feet" is simply the yield per foot multiplied by 100. To customize this for your farm, you would want to multiply yield per foot by your most common row length. Both of these operations are easily done with the computer spreadsheet available from CSA Works, and if you don't have a computer, we can customize the plan for your farm.

Crop	*Per share per year average*	*Yield per row foot*	*Annual goal per 100 shares in pounds*	*Yield per 100 row feet*	*Rows needed @ 100 feet per row*	*Percentage of total acreage*	*No. of weeks harvested*
Mizuna	3.37	0.57	337	57	5.92	0.40	20
Mustard greens	1.93	0.57	193	57	3.39	0.23	21
Onions	8.00	0.70	800	70	11.45	0.77	0
Parsley	1.10	0.50	110	50	2.20	0.22	17
Parsnips	3.00	0.48	300	48	6.21	0.63	1
Peas	2.00	0.20	200	20	10.00	2.02	2
Hot peppers	1.50	0.76	150	76	1.97	0.40	10
Sweet peppers	10.51	1.04	1,051	104	10.14	2.05	10
Potatoes	30.00	0.80	3,000	80	37.50	7.57	4
Pumpkins	9.88	1.40	988	140	7.06	2.56	2
Radishes	2.62	0.33	262	33	7.95	0.53	12
Rutabagas	4.16	1.40	416	140	2.97	0.30	4
Scallions	5.60	0.40	550	40	13.75	0.92	14
Acorn squash	6.44	1.20	644	120	5.37	1.95	5
Blue Hubbard squash	1.73	1.40	173	140	1.23	0.45	1
Buttercup squash	2.44	1.66	244	166	1.47	0.53	3
Butternut squash	13.14	2.00	1,314	200	6.57	2.39	4
Spaghetti squash	4.90	1.20	490	120	4.08	1.48	3
Summer squash	19.00	5.18	1,900	518	3.67	1.33	13
Sweet Dumpling squash	1.00	0.50	100	50	2.00	0.73	1
Delicata squash	2.45	0.50	245	50	4.90	1.78	3
Tatsoi	3.00	0.20	300	20	15.02	1.01	18
Tomatoes	36.59	5.03	3,659	503	7.27	1.47	9
Tomatoes, plum	9.00	2.50	900	250	3.60	0.73	4
Turnips	1.83	0.33	183	33	5.50	0.56	6
Watermelon	24.95	7.09	2,495	709	3.52	1.28	4
TOTAL	**405.30**		**51,273.96**		**686.00**	**100.00**	

Source: © 1997 CSA Works.

Planning for Contingency

As the Colorado potato beetle has proven a few times over, 660 row-feet of eggplant does not equal 1,000 pounds of baba ganoush. The actual crop amounts listed on the chart assume an ideal world. You should add in a contingency of 5 to 25 percent, depending on how important the crop is to your members. You don't want to run out of staples like lettuce, corn, tomatoes, or carrots—so your contingency for these crops might be close to 25 percent. Smaller margins will work for the lesser crops. If you run out of fennel or kohlrabi, you are not likely to cause a riot.

spreadsheet with additional columns for dressing applied, BD sprays 500 and 501, weather when sown, and field or bed number. Once set up on a computer or simply copied, charts like these are easy to use.

Biodynamic growers use a zodiac calendar, available from the Biodynamic Association under the title *Stella Natura: Inspiration and Practical Advice for Home Gardeners and Professional Gardeners in Working with Cosmic Rhythms* (see CSA Resources for how to acquire). They provide a guide to the optimum planting and harvesting times for leaf, flower, fruit, and root crops according to events in the solar system. The calendar provides an explanation of its use. Some growers swear by this method. Anne Mendenhall, a seasoned Biodynamic farmer, points out that real field conditions do not always accommodate the cosmic rhythms, so growers should not feel too upset if they cannot follow this scheme to perfection.

While CSAs stress seasonal eating, most CSA farms try to extend the growing season in order to provide desirable variety in the shares and to increase the amount of food members purchase from the farm. The added cost of season extension needs to be factored into share prices. Farm notes in the CSA newsletter are a good place to educate members about the care and labor involved in babying along

WHAT IS A SHARE

The chart below shows what we estimate to constitute a share. Shareholders share the risks and the bounties inherent in growing. As you can see here, yields fluctuated according to yearly weather variations. Weights are in pounds.

Vegetable	*Est. '94*	*Actual '94*	*Est. '95*	*Actual '95*	*Est. '96*
Apples	—	—	10.0	10.0	10.0
Asparagus	2.0	2.1	2.0	3.0	2.0
Basil	1.0	4.5	2.0	3.0	2.0
Beans	20.0	21.0	20.0	18.0	20.0
Beans, dry	2.0	2.0	2.0	—	2.0
Beets and greens	15.0	17.0	15.0	6.8	15.0
Broccoli	10.0	11.8	10.0	19.0	10.0
Brussel sprouts	2.0	3.0	2.0	—	2.0
Cabbage	15.0	9.0	15.0	6.6	15.0
Celeriac	—	—	—	0.5	0.5
Chinese cabbage	4.0	5.0	4.0	5.7	4.0
Carrots	30.0	24.0	30.0	30.0	30.0
Cauliflower	5.0	10.5	5.0	4.5	5.0
Cucumber	15.0	25.5	15.0	13.0	15.0
Daikon	2.0	0.4	2.0	2.0	2.0
Eggplant	3.0	0.0	3.0	2.2	3.0
Flowers	pick your own		pick your own		
Garlic	—	—	—	0.25	0.25
Herbs	pick your own		pick your own		
Kale	5.0	2.5	2.0	2.4	2.0
Kohlrabi	2.0	—	2.0	—	2.0
Leeks	1.0	2.0	1.0	3.3	1.0
Lettuce	20.0	12.0	20.0	23.0	20.0
Melon	20.0	15.0	20.0	29.0	20.0
Onions	15.0	4.3	15.0	15.5	15.0
Parsley	1.0	1.5	1.0	1.5	1.0
Parsnips	—	—	—	1.0	1.0
Peas	5.0	40.0	5.0	7.5	5.0
Snap peas	2.0	2.3	2.0	1.0	2.0
Snow peas	2.0	0.5	2.0	2.0	2.0
Peppers	2.0	0.8	3.0	2.0	3.0
Potatoes	50.0	14.5	50.0	36.0	50.0
Pumpkin	30.0	25.0	25.0	10.0	25.0
Radish	5.0	1.5	5.0	2.0	5.0
Rhubarb	—	—	—	2.6	2.5
Rutabaga	10.0	0.0	10.0	10.0	10.0
Scallions	4.0	2.9	4.0	1.5	4.0
Spinach	2.0	3.4	2.0	1.5	2.0
Winter squash	30.0	15.0	25.0	5.7	25.0
Summer squash	15.0	31.0	20.0	8.0	20.0
Swiss chard	5.0	2.5	5.0	3.0	5.0
Tomatoes	40.0	21.6	40.0	103.0	40.0
Turnip	4.0	3.3	4.0	9.0	4.0
Watermelon	4.0	5.0	4.0	14.0	4.0
Zucchini	10.0	—	—	20.0	20.0

Source: Willow Pond Farm CSA.

HARVEST CALENDAR AND ESTIMATED ANNUAL AMOUNTS

CROP	ANNUAL AMT	MAY	JUNE	JULY	AUG	SEPT	OCT	NOV, - DEC,
ASPARAGUS	5 #							
RHUBARB	5 #							
LEAF LETTUCE	6 - 8 heads							
WASHED SALAD MIX - 6 oz. bags	20 bags							
RADISHES **	2 bunches							
SPINACH - 8 oz. bags	16 bags							
PARSLEY, CILANTRO**	6 bunches ea.							
PEAS; SNAP AND SNOW	2 # each							
SAUTE` GREENS ** 8 oz. bags	8 bags							
CHARD & OTHER COOKING GREENS	9 bunches							
STRAWBERRIES	6 qt.							
FENNEL **	4							
BEETS	15 #							
BROCCOLI	15 #							
BEANS; FRENCH PETITE	10 #							
ZUCCHINI / SUMMER SQUASH	40							
CUCUMBER	20							
GREEN ONIONS	6 bunches							
CARROTS	30 #							
POTATOES	25 #							
GARLIC	30 bulbs							
SWEET CORN	3 doz.							
TOMATO / CHERRY TOMATOES	20 #							
PEPPERS; SWEET / HOT	75 / 25							
BASIL **	2 #							
ONIONS, RED , YELLOW	30 #							
CELERIAC	6							
CABBAGE; GREEN, RED, NAPA	6							
MELONS	10							
RASPBERRIES	4 pt.							
RUTABAGA	5 #							
TURNIP	4 #							
DAIKON /WINTER RADISH **	3 #.							
PARSNIP	4 #							
CAULIFLOWER	5 heads							
BRUSSELS SPROUTS	4 #							
WINTER SQUASH	15							
KALE; GREEN, FLOWERING **	6 bunches							
JERUSALEM ARTICHOKE **	3 #							
SAGE, OREGANO, THYME**	2 bunches ea.							
EGGPLANT	6 - 8							
SWEET POTATOES	6 - 8 #							
HORSERADISH	1 #							
GROUND CHERRY **	1 Pint							

** Starred items are available to be added to your box as desired but are not regularly included in the weekly share.

SUCCESSION PLANTING PLANNING CHART

Crop	*Variety*	*Greenhouse seeding date*	*Direct seeding date*	*Area/ Lin Ft*	*Seeds needed*	*#Plts (PL/SD)*	*Plant (#Bed) #Trays/type*	*Harv/Share*	*Harv/WS*
Arugula			4-19, 5-6, 5-23, 6-11, 7-10, 8-2	2 beds/1,200	12 oz.		6 (1 row)	24 bunch	
Beans	Jumbo		4-30, 5-17, 6-6, 7-1, 7-29	10 beds/4,000	25 lb.		5 (2 rows)	20 lb.	
	Windsor Fava		4-1	2 beds/800	5 lb.		1 (2 beds)	2 lb.	
Beets	Chioggia		4-14	4 beds/2,400	3 lb.		1 (4 beds)	25 lb.	
	Formanova		5-11	2 beds/1,200			1 (2 beds)		
	Lutz Green Leaf		6-4	4 beds/2,400			1 (4 beds)		25 lb.
Bok choy	Mei Qing Choy	8-2		1 bed/600	¼ oz.	600	5/128	5 heads	
Broccoli	Green Valient	6-30		4 beds/1,600	3 oz.	2,700	13/128	10 heads	
	Emperor	7-20		6 beds/2,400			19/128		
Brussel sprouts	Prince Marvel	5-16		2 beds/800	¼ oz.	600	5/128	10 lb.	10 lb.
Burdock	Takinogawa		4-13	1 bed/600	¼ lb.		1 (1 bed)	10 lb.	
Cabbage	Early Jersey Wake	4-3, 5-16, 6-12		9 beds/3,600	½ lb.	2,400 (800)	7/128	12 heads	
	Storage #4	6-12		2 beds/800	1,000 seeds	500	4/128		10 heads
Chinese cabbage	Blues	8-2		1 bed/600	¼ oz.	600	5/128	3 heads	
Carrots	Thumbelina		4-4	6 beds/3,600	2 lb.		1 (6 beds)	50 lb.	50 lb.
	Narova		5-1	4 beds/2,400			1 (4 beds)		
	Bolero		6-13	6 beds/3,600			1 (6 beds)		
Cauliflower	Violet Queen	5-16, 6-12		8 beds/3,200	12 gr.	2,000	8/128	10 heads	3 heads
	Amazing					(1,000)			
Celery	Ventura	3-29		2 beds/1,200	¼ oz.	900	7/128	6 stalks	
Celeriac	Diamante	3-29		1 bed/600	¼ oz.	450	4/128		10 lb.
Corn	Sugar Buns	4-30	5-17	10 beds/400	20 lb.	2,000	1(5b)+ 6/128	50 ears	
	Kiss 'n Tell		6-5, 22	20 beds/8,000			2 (10 beds)		
Collards	Champion	5-23		½ bed/200		150	2/128	8 bunch	7 bunch
Cucumbers	Marketmore 86		5-26, 6-22	6 beds/1,200	7 oz.		2 (3 beds)	20 lb.	
Eggplant	Rosa Bianca	4-3		2 beds/800	8 gr.	600	9/72	10 lb.	
Fennel	Zefa Fino		5-3, 6-2	1 bed/600	2 oz.		2 (½ bed)	8 stalks	
Garlic	French Red		Mid-October	5 beds/3000	90 lb.		(5 beds)	30 bulbs	15 bulbs
Greens	Spring Raab		4-19, 5-6, 5-23, 6-11, 7-10, 8-2	12 beds/7200	25 lb.		6 (2 beds)	24 bunch	
Herbs	Sacred Basil	4-3			¼ oz.	400	4/128	18 bunch	
	Borage	4-7				150	2/72		
	Cilantro		5-3, 6-2	⅔ bed/400	½ oz.		2 (⅓ bed)	10 bunch	
	Fern Leaf Dill		5-3, 6-2	⅔ bed/400	8 gr.		2 (⅓ bed)	10 bunch	
	Garlic Chives	4-3			pkt.	72	1/72	10 bunch	
	Tarragon	4-3			pkt.	72	1/72		
	Sorrel	4-3			pkt.	72	1/72	10 bunch	
Kohlrabi	Kolibri		6-19	½ bed/300	1 oz.		1(½ bed)	10 lb.	

Source: Quail Hill Farm

early crops or keeping late crops from freezing. A CSA farmer in northern Vermont uses her skill at producing early tomatoes in a greenhouse as a way of attracting new members. She displays the tomatoes at her farm stand, but when people want to buy, she informs them that the quantity is limited so only members of her CSA enjoy the privilege. Signs them right up!

An alternative to growing all the crops for a CSA on one farm is for several farms to cooperate. Each farm can produce the crops it does best and the production system can be simpler. On the other hand, a cooperative venture presents complexities of a different order: the need to agree on quality control, scheduling, fairness in allocating crops, and dividing the proceeds. (See chapter 19 for more on this topic.)

ANIMALS

by Robyn Van En

There are several ways that a group of sharing members can obtain meat and poultry products. The first choice may be the primary farm site that is already providing the vegetables for the shares. The second may be to connect with another farm that is already producing these items or is in a position to start livestock production. The "Resource Guide for Producers and Organizers" from Iowa State University Extension has a useful section on "CSA Livestock and Animal Products," which outlines some of the regulatory complexities of selling animal products in Iowa. Regulations may vary from state to state.

A primary goal of CSA is to develop each site to its most diverse and sustainable potential, so animals invariably come into that equation. Introducing animals can expand the product selection for members beyond vegetables, while providing the farm with an accountable and consistent source of manure for the compost piles. You do not *have* to use the animals for other products; they may, instead, be "farm pets," who help out by building up the biological activity on the grassland through their foraging and manuring.

Biodynamic farmers believe that the animals define the farm as an organism. Trauger Groh writes, "The exchange of the feeds from the farm organism to the herds, and from the herd's manure back to its soil, creates a process of mutual adaptation. . . . The aim and the result of this local interchange between animal and forage is permanent fertility" (*Farms of Tomorrow Revisited*, p. 24).

Whether you adhere to Biodynamic teachings or not, developing a site to its most sustainable potential requires recycling nutrients. The ideal situation is to create a complete system on site where you do not have to import any of the necessary inputs. This entails the production of all feed for the animals at the site, and then use of all of their waste products in the compost to reincorporate the nutrients derived from the fields back into the soil and the crops once again.

While this has been traditional practice at farms and gardens around the world for thousands of years, some recent writings on the subject provide a theoretical understanding of the processes involved. I highly recommend Alan Savory's work on Holistic Resource Management, Robert Rodale's writings on Regenerative Agriculture, and Rudolf Steiner's essays on agriculture.

If a CSA farm does not have animals already, the core group needs to discuss the various options and pros and cons to decide which animals are appropriate and to determine the member demand for animal products. I recommend extensive research, especially if animal husbandry is a first-time venture for the production crew. Since all members may not be interested in shares of meat or eggs, the core may want to create a separate budget and share price for these items. Some CSAs include eggs and meat in the regular shares, others do not. Silver Creek Farm offers shares of the lamb, chicken, and eggs they raise themselves, as well as beef and goat cheese, which neighboring farms deliver. Supplying other products is another way to get the larger community involved with your CSA. Your community may have lots of home gardeners who do not want a vegetable share, but may be interested in eggs or meat from free-range animals that have been raised organically, treated humanely, and not dosed with antibiotics or hormones.

The presence of animals at a CSA site may be a moral, religious, or philosophical issue for some of your members. While only 2 percent of the population are vegetarians, the percentage among CSA sharers is probably a lot higher. Members of one CSA took exception to the use of animal byproducts as soil amendments and wildlife deterrents, arguing that their use helps to perpetuate the livestock industry. The core group acknowledged their concern, but decided to overrule their objection. The majority felt that the needs of the soil and crop protection were more important, and that integrating a sustainable number of animals into the farm and raising them in a humane way could not be equated with supporting the industrialized livestock industry. The discussion provided a good occasion for education for all of the members on the difference between humane animal husbandry and industrial meat production.

Even if a CSA is based on a corner lot in an urban center, some animal waste products can be incorporated into the compost piles to achieve optimum microbiological activity, and to build up nutrient levels in the soil. At an urban site, you might not be able to keep animals, but you can buy bags of dehydrated manure from garden supply stores. In a few cities, urban gardeners have been able to gain the cooperation of the mounted police, who were happy to have a good use for the cleanout from their stables. A small amount of manure goes a long way when mixed with the plant wastes from the garden as the basis for the compost, and, ideally, members could return their kitchen and yard scraps to add to the piles. Quail Hill Farm has been exploring community-supported composting, supplying members with biodegradable bags, designed by Woods End Lab, for their food scraps. (See chapter 22 for more details.)

Rabbits are the most suitable animals for an urban site. They provide excellent manure for a contained nutrient cycle. The site would have to be large enough to grow their food and to give them room to run around. Security could also be a problem.

Animals are a lot of work and a daily responsibility. Even to board or allow animals at a CSA site for their company and pasture management reflects upon the whole project. The CSA group must be sure that the animal owner complies with local health and animal maintenance codes, and that insurance coverage acknowledges the animals' presence and insures accordingly for everyone's protection.

River Brook Farm
Cochecton, New York

Most CSAs have members who would be interested in shares of eggs and stewing chickens, so hens may be the first critters a group will consider. Acquiring and outfitting a flock of chickens is still a major step. It is difficult to set up a small-scale chicken operation that combines fair regard for the birds' well-being and relative happiness with cost-effectiveness. The benefits are many: You can't beat the eggs for flavor, freshness, and nutrients, and the litter and waste from the chicken house and yard are excellent in the compost pile. You can expect 112 pounds of manure per year per bird. Covering the costs is another matter. Chicken tractors are a worthwhile contraption, but you must consider the work of construction and the time it takes to move them around the field. Chickens are vulnerable to wild predators and can also be temperamental.

I had always had a flock of mixed-breed chickens, ten to twenty at a time. With the CSA project going and demand for eggs increasing dramatically,

I thought I'd try a bigger flock. Friends leaving town offered me two hundred hens, so we moved them to Indian Line Farm. Once the hens and roosters settled into their new "digs," free-ranging with organic grains, they all seemed happy and healthy, laying eleven dozen eggs a day. Until I went away for two days, during which time they missed some component of their feed ration and stopped laying for an entire month! Of course, they continued eating, but we got less than a dozen eggs from two hundred hens. We could not supply the accounts we had established for eight dozen eggs a day. I'm sure that part of the problem was that the birds were mostly hybrids and temperamental. Our plan was to build up to mixed breeds over time. But along with the coon and coyote problems and the rate the flock ate down their ranging area, the chicken business proved to be way more trouble than it was worth to us. Since our experience, I've read similar stories, so I can't stress enough the need to research and talk to experienced folks *before* you begin. You do not want to compromise your own energy reserves or the existing CSA during your learning process.

Preliminary research may determine that your production crew or CSA site is not prepared or appropriate for having animals. Holistic Resource Management training in ecological decision-making is very helpful in figuring out what steps you can take to have animals in the future. In the meantime, you can pursue a relationship with local animal farmers, who can supply your members with an expanded product selection and your compost piles with manure. When CSA projects start networking with other farms the possibilities are myriad. CSA has the capacity to involve a wide variety of farm operations in supplying consumers with a broad selection of regionally produced products. Transcontinental and intercontinental relationships of product exchange among sustainable producers are within the realm of possibility.

IN A WORLD OF GROWING POPULATION and shrinking fertile land, CSAs, even those with relatively unskilled farmers, have proven their capacity to produce enough food for twenty families or more on each acre. As CSA farms mature, their production becomes more intense, whether on one farm or several associated farms. Where industrialized agriculture seems to have passed its peak of productivity, and more chemicals no longer means more output, biological farms with community support offer long-term prospects of unlimited promise.

HANDLING THE HARVEST 14

Suitable temperature and moisture, and the separate storage of incompatible types of produce are the keys to keeping fruits and vegetables fresh and attractive as long as possible.

—TRACY FRISCH, "HOW TO KEEP FRESH FRUITS AND VEGETABLES LONGER WITH LESS SPOILAGE"

WITH SHARERS OR WITHOUT, the harvest crew needs certain basic tools: sharp knives, clippers, rubber bands for bunching, containers to pick into, a cart, wagon, or truck to haul the food in from the fields, a setup for washing the produce in clean, cold water, and a cool place to store it once picked. Harvesting and post-harvest handling take up 50 percent or more of the time and energy involved in the production of fruits and vegetables. This is the underside of vegetable production, which we need to make more efficient. Michael Docter and Linda Hildebrand have made a video to demonstrate their techniques for harvesting 6,000 pounds of produce with three people, a flatbed truck, and a tractor (available from CSA Works, see address in CSA Resources). We also need to understand a lot more about how to maintain quality after harvest.

Different crops have differing storage requirements. Tracy Frisch's little booklet, "How to Keep Fresh Fruits and Vegetables Longer with Less Spoilage," is an excellent guide to optimum storage temperatures for maintaining produce quality and nutritional value. If kept from obvious wilting, many vegetables will still *look* all right after several hours of storage at the wrong temperature, but they will have lost a significant portion of their nutritional value and storage life. Watershed Organic Farm CSA provides members with a handbook that summarizes briefly how to store the vegetables to retain freshness and nutritive content. The GVOCSA and MACSAC CSA food/cookbooks contain this information too. (See CSA Resources for a selection of publications.)

An Extension newsletter on direct marketing gives this sensible advice on post-harvest handling:

> In general, a fruit or vegetable is at its prime eating quality when it is harvested (the few exceptions to this are fruits or melons that *may* be ripened after harvest). This does not necessarily mean

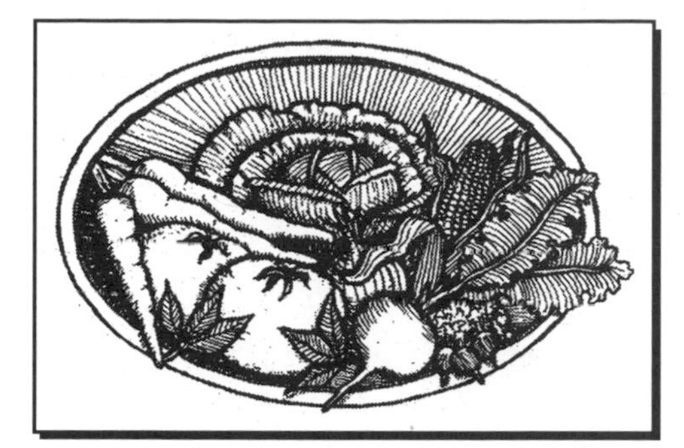

Washershed Organic Farm CSA
Pennington, New Jersey

> that a vegetable would not be better if it were left on the plant longer, but simply that it does not continue to improve after it is removed from the mother plant. In fact, the quality of that item a few hours, days, or weeks later will depend on how well the rate of deterioration is controlled. From a practical standpoint, this means controlling the storage or display environment, especially temperature and humidity, and thus reducing deterioration due to respiration and transpiration (water loss). (Cornell, Department of Agricultural Economics, #10, 1979)

There is a lot of information on crop standards and post-harvest handling on the Internet, accessible through www.agnic.org/agbd/, the National Agricultural Library's listing of agricultural resources.

The goal of this chapter is not to teach you how to harvest crops (only hands-on experience can really do that), but rather to raise some important considerations.

All the studies of CSA members show that access to high-quality, organic, fresh, and locally grown produce is at the top of the list of reasons for joining. (See Master's theses by Dorothy Suput, 1992, and Tim Laird, 1995; and Rochelle Kelvin's Rodale survey, 1994.) Gerry Cohn at UC Davis surveyed the members of ten CSAs in California in 1995. He concluded that "the three most important aspects in the decision to join a CSA are: (1) fresh, seasonal produce; (2) environmental concerns; and (3) supporting a local farmer. The three least important aspects are: (1) only source of organic produce; (2) education about farming; and (3) convenience. It is also notable that price is not considered particularly important" (Gerry Cohn, "Community Supported Agriculture"). In Vermont, Jane Kolodinsky found price was a more important factor ("An Economic Analysis of Community Supported Agriculture Consumers").

Since freshness and quality are so important to CSA sharers, harvesting must be designed accordingly. Greens must be cut while the day is still cool, never left in the sun, washed, and stored in a cool place as quickly as possible. Most vegetables left sitting in a warm place drop drastically in nutritional value within minutes. (See the chart on p. 146–47 for information on plant respiration, an indicator of quality loss during storage.) You may not need expensive hydrocooling, but you must pay attention to detail. As soon as you can afford refrigeration, make the investment. It will repay you in savings of nutritional values, in convenience, and in improvement to the storage life of your vegetables. The walk-in cooler at Rose Valley freed us to pick some things the day before distribution and lowered my blood pressure.

Plant Respiration

The energy that plants store through photosynthesis is released by *cellular respiration,* both while a plant is living, and after it has been harvested. Of course, energy that is lost through respiration is no longer available to us when we consume the food. Therefore, the greater the rate of respiration post-harvest, the faster a fruit or vegetable loses nutritional quality. The chart on p. 146–47 will help you understand what happens to your crops at various temperatures. For example, at 60 degrees Fahrenheit, bunched carrots respire more rapidly than do topped carrots. Heat is more damaging to some crops than others.

RESPIRATION RATES OF FRUITS AND VEGETABLES
(expressed as rate of carbon dioxide production [mg/kg/h] at various temperatures)

	Temperature					
Commodity	**0°C (32°F)**	**4–5°C (40–41°F)**	**10°C (50°F)**	**15–16°C (59–60°F)**	**20–21°C (68–70°F)**	**25–27°C (77–80°F)**
Apples, summer	3–6	5–11	14–20	18–31	20–41	—
Apples, fall	2–4	5–7	7–10	9–20	15–25	
Apricots	5–6	6–8	11–19	21–34	29–52	—
Artichokes, globe	14–45	26–60	55–98	76–145	135–233	145–300
Asparagus	27–80	55–136	90–304	160–327	275–500	500–600
Avocados	—	20–30	—	62–157	74–347	118–428
Bananas, green	—	—	—	21–23	33–35	—
Bananas, ripening	—	—	21–39	25–75	33–142	50–245
Beans, lima	10–30	20–36	—	100–125	133–179	—
Beans, snap	20	35	58	93	130	193
Bean sprouts	21–25	42	93–99	—	—	—
Beets, topped	5–7	9–10	12–14	17–23	—	—
Beets, with leaves	11	14	22	25	40	—
Berries						
Blackberries	18–20	31–41	62	75	155	—
Blueberries	2–10	9–12	23–35	34–62	52–87	78–124
Cranberries	—	4–5	—	—	11–18	—
Gooseberries	5–7	8–16	12–32	27–69	41–105	—
Raspberries	18–25	31–39	28–55	82–101	—	—
Strawberries	12–18	16–23	49–95	71–92	102–196	169–211
Broccoli	19–21	32–37	75–87	161–186	278–320	—
Brussels sprouts	10–30	22–48	63–84	64–136	86–190	—
Cabbage	4–6	9–12	17–19	20–32	28–49	49–63
Carrots, topped	10–20	13–26	20–42	26–54	46–95	—
Carrots, bunched	18–35	25–51	32–62	55–106	87–121	—
Cauliflower	16–19	19–22	32–36	43–49	75–86	84–140
Celery	5–7	9–11	24	30–37	64	—
Celeriac	7	15	25	39	50	—
Cherries, sweet	4–5	10–14	—	25–45	28–32	—
Cherries, sour	6–13	13	—	27–40	39–50	53–71
Citrus						
Grapefruit	—	—	7–9	10–18	13–26	19
Lemons	—	—	11	10–23	19–25	20–28
Limes, Tahiti	—	—	—	6–10	7–19	15–45
Oranges	2–5	4–7	6–9	13–24	22–34	25–40
Cucumbers	—	—	23–29	24–33	14–48	19–55
Endive	45	52	73	100	133	200
Figs, fresh	—	11–13	22–23	49–63	57–95	85–106
Garlic	4–14	9–33	9–10	14–29	13–25	—
Grapes, American	3	5	8	16	33	39
Kale	16–27	34–47	72–84	120–155	186–265	—
Kohlrabi	10	16	31	49	—	—
Kiwifruit	3	6	12	—	16–22	—
Leeks	10–20	20–29	50–70	75–117	110	107–119
Lettuce, head	6–17	13–20	21–40	32–45	51–60	73–91
Lettuce, leaf	19–27	24–35	32–46	51–74	82–119	120–173
Lychees	—	—	—	—	—	75–128

	Temperature					
Commodity	**0°C (32°F)**	**4–5°C (40–41°F)**	**10°C (50°F)**	**15–16°C (59–60°F)**	**20–21°C (68–70°F)**	**25–27°C (77–80°F)**
Mangoes	—	10–22	—	45	75–151	120
Melons						
Cantaloupes	5–6	9–10	14–16	34–39	45–65	62–71
Honey dew	—	3–5	7–9	12–16	20–27	26–35
Watermelons	—	3–4	6–9	—	17–25	—
Mushrooms	28–44	71	100	—	264–316	—
Onions, dry	3	3–4	7–8	10–11	14–19	27–29
Onions, green	10–32	17–39	36–62	66–115	79–178	98–210
Okra	—	52–59	86–95	138–153	248–274	328–362
Olives	—	—	—	27–66	40–105	56–128
Papayas	4–6	—	—	15–22	—	39–88
Parsley	30–40	53–76	85–164	144–184	196–225	291–324
Parsnips	8–15	9–18	20–26	32–46	—	—
Peaches	4–6	6–9	16	33–42	59–102	81–122
Pears, Bartlett	3–7	5–10	8–21	15–60	30–70	—
Pears, Kieffer	2	—	—	11–24	15–28	20–29
Peas, unshelled	30–47	55–76	68–117	179–202	245–361	343–377
Peas, shelled	47–75	79–97	—	—	349–556	—
Peppers, sweet	—	10	14	23	44	55
Persimmon, Japanese	—	6	—	12–14	20–24	29–40
Pineapples, mature-green	—	2	4–7	10–16	19–29	28–43
Plums, Wickson	2–3	4–9	7–11	12	18–26	28–71
Potatoes, immature	—	12	114–121	14–31	18–45	—
Potatoes, mature	—	3–9	7–10	6–12	8–16	—
Radishes, with tops	14–17	19–21	31–36	70–78	124–136	158–193
Radishes, topped	3–9	6–13	15–16	22–42	44–58	60–89
Rhubarb, stalk	9–13	11–18	25	31–48	40–57	—
Romaine	—	18–23	31–40	39–50	60–77	95–121
Rutabagas	2–6	5–10	15	11–28	41	—
Spinach	19–22	35–58	82–138	134–223	172–287	—
Squash, butternut	—	—	—	—	—	66–121
Squash, summer	12–13	14–19	34–36	75–90	85–97	—
Sweet corn, with husks	30–51	43–83	104–120	151–175	268–311	282–435
Sweet potatoes, uncured	—	—	—	29	—	54–73
Sweet potatoes, cured	—	—	14	20–24	—	—
Tomatoes, mature-green	—	5–8	12–18	16–28	28–41	35–51
Tomatoes, ripening	—	—	13–16	24–29	24–44	30–52
Turnips, topped	6–9	10	13–19	21–24	24–25	—
Watercress	15–26	44–49	91–121	136–205	302–348	348–438

Note: Respiration rate often shows as a range. To compute heat evaluation rates at harvest time, use either the highest figure or median one. To convert to Btu/ton (2,000 lb.)/24 hour day, multiply respiration rate by 220. To convert to kcal per 1,000 kg/24 hour, multiply by 61.2. Some data included for low temperatures that cause injury to certain commodities or cultivars, such as avocado, mango, okra, papaya, pepper, pineapple, tomato, and zucchini squash; these low temperatures are potentially dangerous and should be avoided.

Source: From R. E. Hardenburg, A. E. Watada, and C. Y. Wang, *The Commercial Storage of Fruits, Vegetables, and Florist and Nursery Stocks,* Agriculture Handbook No. 66, USDA, 1986. With permission.

CLEMENS KALISCHER

Washing kale at Green City Farm, Burlington, Vermont.

You must make your own decisions about quality standards. As Robyn put it: "Many of the current CSA farmers come from an extensive farmers' market or restaurant supply background, where providing a good-looking product was integral to their success, but required a lot of prep and groom time. Often these farmers still strive to maintain high standards of appearance, even though the extra work can impact the share price. Other farmers do the absolute minimum, removing dead leaves and weeds, sloshing the carrots and radishes in a bucket of clean water to rinse off the dirt clods, spritzing the lettuce with water and storing it in the shade or a cool place to deter wilting, and removing the obvious slug. The results may not look like supermarket produce, but the quality, variety, and nutritional value cannot be beat.

CSAs can certainly work to set themselves apart by the aesthetically perfect appearance of their produce, but their budget will rarely cover the cost of the labor required. Pound for pound, CSA members get a tremendous return on their investment, even if they have to do an extra bit of washing or separate the radishes from the string beans in the bag. If this is a real issue with the members, discuss it at your annual meeting or ask for comments in the newsletter or a year-end survey. Better yet, get members involved in distribution, and see what standards they can maintain in the rush to make the shares look as nice as possible in the shortest amount of time.

"CSA is supposed to be socially responsible, meaning that all hours of labor should be paid for. Little measures such as not scrubbing the carrots, rubber-banding the herbs, or putting several items in one bag will save an amazing amount of time and materials. Understandably, this goes totally against the training or standards of some growers.

"Members who participate in the harvesting are less likely to complain about dirt or bugs. We had a member who loved CSA and had an infant son who was allergic to almost everything, other than vegetables and fruit, so each week she would cook and purée almost an entire share for him. At one point, though, she couldn't refrain from telling us that she was annoyed with the mud puddle she'd get in her kitchen each month when she dealt with her winter share. We explained that crops stored in a root cellar must be left with the soil on to retain their moisture content, and invited her to help with the winter distribution.

"She did show up to help. Eight hours at 35 degrees, handling 1,500 pounds of vegetables, wearing as many clothes as she could while still being able to move, taught her a lot. It was significantly colder outside the root cellar, and the thought of how many more hours it would take to wash and dry everything in those conditions made her realize how much she had taken for granted."

Finally, though, your standards are up to you and your members. Robyn and I argued about the name for this book. She wanted to call it, "Everybody Gets Some Crooked Carrots." At Rose Valley, we did not include crooked carrots in the regular shares, but sold them in bulk as for juicing. Our CSA members are mainly city people, described by core member Pat Mannix as, "recovering Wegman's addicts" (Wegman's is the local super-supermarket). We try to meet their expectations more than halfway. To get Greg Palmer, who can't throw anything away, and a few of the members who are the same way, to adhere to my standards, I urge them to carry a special bag while harvesting. They get to keep the vegetables they just *have* to pick, although they know I would reject them. Our own kitchen, the local soup kitchens, and the food bank are happy to take the culls.

◆

PICKING AND PACKING is the forgotten end of vegetable production. Low-paid migrant farmworkers do most of this work. If we ever hope to have a just and sustainable food system, we must face this reality and figure out how to change it. Inviting the members of CSAs to help harvest their own food is a good beginning.

DISTRIBUTING THE HARVEST 15

As raw materials production is moved around the world to where land and labor are cheapest, local economics and ecologies are destroyed for economic reasons.

—JOAN GUSSOW, *CHICKEN LITTLE, TOMATO SAUCE AND AGRICULTURE*

NO TWO CSAs DISTRIBUTE their shares in exactly the same way. The location of the farm, the distance from sharers, storage facilities, labor and trucking availability, and the amount of member participation all affect the distribution system. Here we will describe the very different and imaginative systems which several CSAs have adopted.

FARM PICKUP

The simplest way to distribute the shares is to have members come to the farm to pick them up, assuming they live nearby. At Indian Line Farm, about half of the members picked up their shares at the farm. The farm crew, comprising the head gardener, his assistant, and member volunteers, picked the vegetables that morning and packed them for the members. The distribution area was the cool, shaded downstairs room in a big old barn. A few jars of flowers decorated the 4 x 10-foot table built from salvaged lumber that held the shares. A blackboard served as communication center, with a list of what was in the harvest, notices, announcements, and space for members to write comments.

The other half of the sharers picked up their bags at the Berkshire Co-op Market, six miles away in Great Barrington. In Robyn Van En's words, "The Co-op had bought shares in the Apple Project and was very willing to offer a corner in the store for our town drop site, allowing regular Co-op shoppers to coordinate their shopping and introducing other community members to the Co-op for the first time. Early on we discovered that we had a pocket of members all living around the mountain in Sheffield. To cut down on their travel (and traffic at the farm), we provided these members with a list of their names and numbers, and let them work out their own rotation of pickups from the farm."

ILF also sent some shares to Lexington, Massachusetts, located about two and a half hours away.

As Robyn described it: "The first year our delivery cost was factored into their share price. The second year, they worked out a rotation among themselves, and each share-holding member came out to the farm once per season (twenty-two weeks) and each month through the winter. The third year, we made arrangements for the local Berkshire Mountain Bakery to take the shares along with their weekly bread delivery to the Boston area. The Lexington households paid a portion of the bakery's transport costs and filled their own weekly bread orders at the same time.

"Piggy-backing with the bakery truck was a real breakthrough, but the ultimate would have been a backhauling connection. That is, to identify a truck that delivers something to your community each week, and have that same, previously empty, returning truck haul shares back to your drop point in the urban center. You would need a site manager to help unload and tend the site, but it would be the most responsible way to go, short of bicycle trailers (like those used at the Topinambur CSA near Zurich, Switzerland), to avoid paying for auto insurance and burning more fossil fuels."

Hidden Valley Farm in Alna, Maine, actually does deliver some of its shares by bicycle! The farm collaborates with the Bicycle Works, which transports the farm's Portland shares by bicycles towing covered trailers. Similarly, Jim Gregory's Fresh Aire Delivery and Recycling Service delivered Magic Beanstalk shares in Ames, Iowa, for $2.80 a week during that CSA's first year of operation.

The Indian Line group tried packing the shares into half-bushel baskets, switched to brown paper bags, then to large plastic bags, and finally ended up back with the paper. When they realized they were using three hundred small plastic bags a week for carrots or greens that needed to stay moist, they found two members who were willing to contribute muslin and sewed six 12 x 14-inch bags for each share. "This allowed for two bags to be at the farm for the next harvest, two filled up with the current week's harvest, and two in the laundry or lost in space." They started by writing each member's name on the bag, then found they saved time by keeping a check-off list.

All the sharers in the Food Bank Farm near Amherst, Massachusetts, pick up at the farm. After the first two years of low membership retention, the farm made a concerted effort to learn what its members wanted. Michael Docter reports:

CLEMENS KALISCHER

CSA member bags her share according to the weekly list at Caretaker Farm.

CSA sharer Darrell Wharton designed a bike trailer to carry his vegetables near Edmonton, Canada.

During our third year, we implemented a mix-and-match greens table. We put all the collards, kale, lettuce, mizuna, arugula, radishes, etc., on one table and gave shareholders a plastic bag of a certain size and let them fill their bag with whatever they wanted. If someone did not like greens and just wanted to fill their bag with lettuce, great. Let them eat lettuce.

Members were thrilled with the system. Finally, they had some real choices. . . . Not only were our members happy, but also the farm saved thousands of dollars in harvest labor under the new system. We no longer had to spend time counting heads of lettuce and bunching and tying greens. Distributing by volume meant we no longer needed to produce uniform head sizes. This enabled us to switch to a much less costly lettuce production system. . . .

In August 1995 we broadened this method of distribution to include the majority of crops we raise on the farm. We place the carrots, beets, broccoli, eggplant, peppers, etc., on another table and give members a bag which they can fill to a certain point with the vegetables they like.

At first, we were afraid that everybody would just take carrots. Our fears were unfounded. While carrot consumption did go up about 20 percent, we distributed just about the same amount of other crops as we had under the old system. Some folks took lots of carrots, others took lots of eggplant—but it all averaged out the same in the end. Consumer response was overwhelmingly positive. (*Growing for Market* [Feb. 1996]: 4–5)

Members not only pick up their vegetables at Caretaker Farm in Williamstown, Massachusetts, but they wash them too. Conveniently located near the compost pile is a sink with a supply of rubber gloves, scrub brushes, and aprons. After depositing their vegetable wastes from the previous week, and covering them with dry material from the nearby pile of leaves, they can step up to the sink and rinse this week's takings. This arrangement not only saves work for the farmers, but keeps the kitchen floors of members clean.

FARM STAND AND FARMERS' MARKET PICKUPS

Farms that run farm stands or sell at farmers' markets have found ways to combine choice and convenience for their sharers. After leaving the boxes for his sharers under a tree near his farm stand and watching the produce wilt by late afternoon, Skip Paul had a brainstorm to combine the two enterprises. During its second year, the Wishing Stone Farm CSA in Little Compton, Rhode Island, started offering its members a line of credit at the farm stand. Rather than purchasing a uniform box, members can choose what they want from the stand at special CSA-only prices, with extra savings on crops that are abundant that week. Skip has reduced their discount on regular prices from 20 to 10 percent, and has added five working shares, which cuts the share price from $300 to $100 plus forty hours of work. Members can also charge to their CSA account other products for sale at the stand such as milk, baked goods, cut flowers, or exotics such as bananas. The farm has a certified commercial kitchen in its barn where a special staff cooks and bakes an assortment of products for the stand's deli section. As a result of this new plan, the retention rate for sharers has soared. In 1998, Wishing Stone is adding share pickup at its farmers' market stand in Providence, saving some members a forty-five-minute drive.

Wishing Stone Farm
Little Compton, Rhode Island

The Goranson Farm in Dresden, Maine, invites its customers to pay in advance and receive a discount at their farm stand. Anyone can buy at their stand, but the 175 or so families who pay in advance are making a special commitment. Jan Goranson asks, "why give interest to a bank, when we can give a discount to our customers?" For $97, members receive $100 worth of food, for $190 they get $200, and for $370 they get $400. As at Wishing Stone, members pay up front and then take whatever they want at the stand until they use up their credit. At the end of the season, money sometimes remains in member accounts. For the first few years, the Goransons returned the unspent money. These days, they send a reminder card and, if there is no response, keep the change.

Earthcraft Farm in Indiana sells a cluster of 180 units to the CSA members who pick up at their farmers' market stand, to apportion at their convenience. Members do not seem to worry about leftover units. Earthcraft's 1997 CSA handbook says, "If you only used half your units, but you ate wonderful meals, tried new recipes, stayed healthy, and enjoyed life with the help of Earthcraft Farm produce, then you got from the subscription what we tried to make available through it." The high retention rate suggests that members agree.

IN-TOWN PICKUP

Farms that are located farther from their members need a more elaborate distribution strategy. Most GVOCSA members live in Rochester, New York, a fifty-mile, one-hour drive from Rose Valley Farm. Despite the distance, each share came with the commitment to work three four-hour shifts at the farm helping to pick the vegetables. At the end of their shift, that week's working members transported the produce to the pickup point at a church in the city. They stored the vegetables, packed in large boxes and bags, in a refrigerated cooler we purchased secondhand from a produce store. In 1997, a walk-in cooler, refrigerated by an air conditioner, replaced the dangerously aging box. I wrote out detailed directions to the distributors, so they knew what to put in the shares. Late in the afternoon, the distribution crew, two or three members supervised by a core group member, arranged the produce for pickup. They weighed out and bagged such items as spinach and carrots. Weighing produce is a job many children enjoy, although combining childcare and distribution does not always go smoothly. A distribution captain told me about a mother who came in with a fussy baby in a backpack. The mother took the job of weighing onions. Every time she bent over to pick up more onions to place in the scale, the baby threw the ones already in the scale basket on the floor.

Members had an hour to collect their shares, which they assembled according to the amounts posted. They checked off their names on a list. Lines of various colors through their names signalled that they had bulk to collect, that they should take boxes back to the farm, or that their payments were overdue. If there was something in the share they didn't like, they could leave it or exchange it for something in the share box. They also picked up and paid for any bulk orders from the previous week, and made orders for the next week, a system that sometimes led to confusion. Marianne Simmons recounted this bulk tale:

CLEMENS KALISCHER

Robyn Van En (right) and a helper weigh and bag vegetables at Indian Line Farm.

MARILYN ANDERSON

Elizabeth Henderson helps Melissa Carlson load vegetable shares to take to Rochester for distribution to the Genesee Valley Organic CSA.

One day, member John Garlock came for his share and found 40 pounds of red cabbage waiting for him that his wife Marilyn had not mentioned ordering. Perplexed, he paid for the order and lugged it home. When I got home from distribution that evening, Marilyn called in a panic. "Marianne, why do I have fourteen red cabbages and what will I do with them?" "Sauerkraut?" I suggested lamely, and explained that she had ordered them. The bulk order sheet indeed had her name next to "red cabbage 40 lb." The price for the cabbage was 40 cents a pound. Marilyn, it turned out, had left off the decimal point. The next morning, I was able to sell thirteen heads to St. Joseph's House of Hospitality for their noon meal.

A modified version of this system continued in 1998. The produce came from four different farms, while Greg Palmer, former apprentice turned partner, and I moved production to another piece of land. A special order coordinator now oversees sign-ups and collects payments for products from other farms, such as organic maple syrup, wine, and grape juice. (See chapter 18 for more details). When pickup time is over, a member of the parish takes unclaimed shares to distribute to families in the neighborhood who need the food.

Like the GVOCSA, the farms that deliver to New York City leave the details of distribution in the hands of their core groups. In most cases, the farmers themselves make the deliveries and then hang out during pickup hours for the chance to make

New Truck

I bought a new truck. Refrigerated. From now on, your boxes will be chilled when they arrive.

We were renting trucks to deliver your boxes. Hundreds of dollars a week for trucks that we have to pick up, return—trucks that didn't cool your boxes.

I spent months looking for the right truck—the right size, the right cooling unit, the right condition, the right price. I ended up buying the approximate truck. I'm already in love with it. Bob and I took it to Rockford last night to show it off. We went into restaurants and bars, asking people if they wanted to see our new delivery truck.

"That reefer unit sitting out in the parking lot?" they'd say. They already knew. There's something about a new truck. It just sits there and shines, like a new baby. People can't help but notice.

It's so seldom that I can go around saying, "I bought a new truck today"—twice so far. The other one I bought in '74, just before the floods that year. It was headed for Pakistan, even had a metric speedometer. The salesman told me it had been sitting on the dock, waiting for the ship, and the deal fell through. I used to wonder who in Pakistan was waiting for that truck. Every Pakistani I imagined waiting for it was so beautiful.

The truck hauled thousands of tons of corn, wheat, beans, thousands of hogs. Tons of building stone. Massive wooden beams.

An artist on the farm went to Egypt. She came back and made things black and white. (Did you check out our barn kitchen?) She painted black and white checks on the bed of my truck.

She made shrines: shrines to memories, shrines to souvenirs, shrines to shoes, to toothbrushes, to teeth—intricate assemblages of Scotch tape, cellophane, string, candles, glue. Paper mâché planets bobbed on wire armatures in her three-foot model of the universe. Extruded teeth gleamed.

One fall, she loaded up the truck with her shrines, her universe model. She called it the Traveling Curio Museum. She toured Wisconsin in my grain truck, pulled into small towns, staked triangular black and white flags from the four corners of the truck to the ground, lowered the catwalk. She set the tooth lady mannequin on the sidewalk. "Tell the tooth lady your dreams," the shrine builder implored the small townspeople. "Turn on the little tape recorder. She wants to know your dreams. Then come in." People ascended the ramp. Often looking for paintings of Wisconsin wildlife, which weren't there.

Twenty-three years later, the second new truck stalked me. I called all over Wisconsin and Illinois seeking a reliable used truck. Sometimes, especially in Chicago, the salesman would say, "Nope, had one a few weeks ago (or even a few years ago!). Seems there's one like that out your way. Some deal fell through, and there's one out there in Rockford somewhere." Refrigerated trucks can become legends.

The Rockford salesman called, offering a viewing of the new truck.

I said, "You can bring it out, but I'm not going to buy it. We're poor. We can't afford a new truck. We couldn't get the credit, anyway. In fact, don't even bring it. It's just wasting your time."

The salesman brought out the truck. He had grown up right down the road from us on a hardscrabble farm. I hadn't seem him in thirty-five years. He had the soft manner that I remembered his whole family having.

I was disturbed at how right the truck looked, sitting in our yard. Sometimes things show up, and I know they're just the dense version of a space that's already there. If I look out of the corner of my eyes, I see that thing just plunking itself into an etheric outline, an Akashic unfolding. It's a hard thing to stave off, because it feels so much like destiny.

I bought a four-wheel-drive tractor like that this spring. I noticed it out of the corner of my eye as I was searching for other iron on a three thousand-acre farm that was phasing out of cabbages. I wouldn't look at it directly. I wouldn't get near it. Bob mentioned it later—"The only thing worth thinking

about that we saw today was that four-wheel-drive tractor, and you wouldn't even look at it."

"That thing scared me, Bob. It's too perfect. We shouldn't even think about it. We can't afford it." I bought it, because Angelic Organics already hosted an etheric space for it. That tractor's one of the main reasons we're on top of the work this year.

Farms fail for many reasons. One reason is too much junk propping up the operation—thwarted plans due to breakdowns, thwarted planting schedules due to chasing parts, thwarted deliveries. Another reason is too much debt. I try to balance the junk with the debt. This time I chose the debt; I bought the truck—five years to pay.

I had to sell my Pakistani truck at auction in '83. I tucked a little terracotta angel under its seat the morning of the sale. I knew the truck was supposed to stay on my farm, but I had to bring in every dollar just to keep the farmstead. A burly farmer bought it. He roared off in it through the spring slush.

I was driving by his farm a few years later; there was an auction at his farm that morning—the bank was selling him out. I stopped and checked for the angel under the seat; it was gone. I wasn't farming then, but I bid on the truck, because I knew it belonged on my farm. I was outbid.

A year later, I was leading a workshop on the farm crisis in a local high school. A hundred scared farmers made up the audience, wondering when they would be sold out, or needing to talk about their recent demise. Through a bank of windows behind the farms, I could see a highway. My old truck glided into view and disappeared.

—John Peterson,
Angelic Organics,
Caledonia, Illinois.
Used by permission.

ERNEST MCDANIEL/COURTESY EARTHCRAFT FARM

Signe Waller and a helper staff the Earthcraft Farm table at the Lafayette, Indiana, Farmers' Market, where CSA members can pick up their shares.

contact with members. For the Roxbury CSA, core members take turns meeting the farm truck and helping to unload the vegetables for the shares each week. Then they oversee the pickup hours. Members make up their own packets, with freedom to select from the produce, up to the posted amounts. Signs alert them to the relative scarcity or abundance of the various items. Sarah Milstein reports that people are pretty good about not hogging early-season tomatoes, and that they appreciate the allowance for choice. The core group also handles sales of items such as locally baked bread to members.

Live Power Farm is located even farther away from its San Francisco members than these farms are from New York City. The farm crew packs the vegetables in bulk boxes and links up with a refrigerated truck making the four-hour run to the city. The core group oversees the breakdown into shares at Veritable Vegetable, an organic produce warehouse. Members take turns picking up for established groupings by neighborhood. As an extra benefit, they are able to make bulk purchases of fruit at wholesale prices from the warehouse. Live Power also has drop-off points in towns nearer the farm.

Not all distribution schemes require such a high level of member involvement. Many CSA farm crews do all the bagging or boxing of shares and all the deliveries. Moore Ranch in Carpinteria, California, Angelic Organics in Caledonia, Illinois, Harmony Valley Farm in Viroqua, Wisconsin, and Full Belly Farm in Guida, California, deliver by truck to distant cities. CSA members volunteer their porches or garages as drop-off points. In exchange, they get a reduction in their own share price or even a free share. Harmony Valley members take turns accompanying the farmers on the city end of the deliveries to help with unloading. Tony Ricci, who drives the produce from Green Heron, his Pennsylvania farm, 115 miles to sites around Washington, D.C., says he could be delivering anything. He becomes a deliveryperson. His main concern is to organize the process well using a computerized system, to time his trips to avoid traffic, and to keep from getting tickets. Two of his drop-off sites are in office buildings where he had contacts on the inside, one with the staff of the World Wildlife Fund.

With his characteristic relish of detail and good humor, Bob Bower of Angelic Organics described their delivery system at the 1997 Northeast CSA conference. They have twenty drop-off sites in a loop around the city of Chicago and its suburbs. Each site has a host who, for a $100 reduction in share price, is willing to allow the use of a porch or leave a garage door open, doesn't mind signs being posted, and is congenial with other members. The farm crew selects sites that are shaded and cool, with protection from the elements. A month ahead of the first delivery of the season, the farm sends out a letter to members with clear, concise instructions on the process. Since the farm is two hours away and the farmers few, they have set up a buddy system among the members, "a tree-like structure for building community." The advantages of this distribution system, according to Bob, are the ease it affords for adding more sites, its flexibility, and the fact that no one has to staff the pickup hours. It also allows delivery close to the homes of the members, within a fifteen minute drive. On the downside, it limits share choice possibilities, though they provide a swap box. (Encouraging members to come in early for a swap hour for trading vegetables with one another did not fly.) A survey of Angelic Organics' members convinced the farm to give up plans for changing to a bulk system at a central distribution site. Only 3 percent of the members favored that option and many wrote answers like this one: "We like the present system because we eat things that we might not choose otherwise and end up enjoying."

Bob talks about the box-packing ritual at the farm as the "highlight of the week. It's really fun! We do it with a conveyor belt. We pack over three hundred boxes in less than an hour. We put binfuls of each vegetable along the side of the belt. Heavy vegetables—carrots, squash—go on the bottom of the box. We'll have seven or eight people: one person making boxes, five putting vegetables in them, and one person closing them. Then the box goes up the conveyor onto another conveyor right into the truck. The music's playing. It's really festive. And now the boxes are on the truck and we clean up. It's the culmination of the whole efforts of the week." The driver comes to the farm at 6 A.M. on Tuesday and Saturday mornings to make the eight- to nine-hour round trip with the refrigerated truck. Before leaving, the driver picks up an instruction sheet on what to do at each site, target arrival times, and a check-off list to remind him to collect recycled bags and boxes, and to tidy up. Bob credits *The E-Myth* (see p. 121) as the impetus for devising their packing and delivery procedures.

Bill Brammer of Be Wise Ranch in California uses a similar system for the eight hundred shares he divides among sixteen delivery sites in and around San Diego. With the help of a computer expert, Bill designed a computer program that tracks billing and deliveries. The program makes it easy to allow members to select their own delivery schedule: weekly, biweekly, or whatever, with skips for vacations. It also prints out the harvest list for the four picking days a week. Every member call for a cancellation is double-logged, noted on the phone log, and then recorded onto the share program so that billing is accurate. Bill has thoughts of redesigning his system so that he can also offer home deliveries of individually chosen orders.

All the produce need not come from one farm. Several farms can coordinate deliveries to a central pickup point. In Lawrence, Kansas, eight farms have joined together in the Rolling Prairie Farmers' Alliance. (See chapter 19 for more details.) On Mondays, they converge on the Merc, the local food co-op, to assemble their packets. Members join the subscription service for the twenty-five-week season. While not asked to help run the service or to do farm work, members are invited to events on the farms, tours, flower walks, and potlucks, to get to know their farmers.

HOME DELIVERY

The ultimate in member non-involvement is the home delivery, called in England a "box scheme." The farm crew picks, sorts, boxes, and delivers to the doorsteps of the members. Katy Sweeney of Malven Farm in Smyrna, New York, has been running her CSA this way since 1990. She says her members are too busy to participate in any way, and she prefers the control this gives her over her farm operation. Flying Dog Gardens in Avon, New York, also did home deliveries in the Rochester area for several years for an extra fee. The Organic Kentucky Producers Association (OKPA), a farmer-owned cooperative, similarly charges a fee for home delivery to the members of its CSA in the Lexington area. Tanyard Farm in West Hartford, Vermont, solicits members' help with deliveries to other members as a way to reduce share payments.

Entrepreneurial schemes are springing up around the country that offer home or office box deliveries. Some of these are quasi-CSAs in that they purchase all their produce from a set group of farms, which they list and describe in their promotional literature to give the flavor, if not much of the reality, of a farm connection. Others purchase vegetables and fruit for the boxes at a terminal market. To call this service a CSA would stretch the definition beyond credibility, since these operations lack any connections to specific farms.

ESPECIALLY FOR CSA FARMS that are at some distance from members, getting distribution to work well for everyone involved while maintaining the high nutritional value of the food seems to present as much of a challenge as growing the food. Even for on-farm pickup, there may be tradeoffs between convenience—that is, extended hours of share availability—and food quality—maximum freshness and appropriate refrigeration. CSA sharers and farmers around the country need to keep putting our heads together to solve this one!

THE WEEKLY SHARE 16

We live in a world that has practiced violence for generations—violence to other creatures, violence to the planet, violence to ourselves. Yet in my garden, where I have nurtured a healthy soil-plant community, I see a model of a highly successful, non-violent system where I participate in gentle biological diplomacy rather than war. The garden has more to teach us than just how to grow food.

—ELIOT COLEMAN, *FOUR SEASON HARVEST*

HOW DOES IT FEEL TO HAVE a garden of your very own? You await with anticipation the first tender greens of spring: dandelion greens, lettuce, asparagus, spinach, baby beets. Later come such delicacies as fresh peas, baby carrots, and the earliest tomato. As summer reaches its peak, you enjoy the wonderful abundance of fresh corn, hot weather crops such as tomatoes, eggplant, peppers, summer squash, cucumbers, and beans. Then comes the glut—tomatoes till you are forced by their excess to can them. As the nights cool, fall broccoli and cauliflower, sweeter carrots, onions, leeks, potatoes, and winter squash warm your tummy. And finally come the crops of late fall: rutabagas and turnips, Brussels sprouts, kale, parsnips, and pumpkins. Time to squirrel food away for the winter. This is what the CSA experience provides, without *all* the work.

Modern supermarkets boast of their ability to supply any food you want all year-round, if you can afford it. As a result, the post World War II generations of consumers have completely lost touch with eating fresh foods in season. When recruiting new members to a CSA, you have to be very clear in explaining what the concept of "eating within the seasons" really means. We try hard to do this at the orientation sessions of the Genesee Valley Organic CSA, and yet, every few years, someone joins only to quit before the end of June because there are no tomatoes in the share.

Biodynamic teachings offer good health reasons for eating what is seasonal for your locality. The brochure for Union Agricultural Institute CSA, in Blairsville, Georgia, presents this Biodynamic wisdom succinctly and refers members to *Louise's Leaves* for recipes (see the CSA Resources section):

> Seasonally, spring greens such as spinach, lettuce, nettles, and poke cleanse and get you going for

the summer. Eat them one way or another every day when you have them. It needn't be monotonous. . . . Almost too soon you'll move along to something else.

Summer fruits such as cucumbers, tomatoes, sweet corn, and peppers cool and give strength. Eat them like there's no tomorrow.

Fall greens like mustards and rape taper you back off, and roots like turnips and radishes give that inner heat and nerve power for loving winter. Stuff yourself to health.

Union Agricultural Institute
Blairsville, Georgia

What and how much should you put in the weekly share? No one has found the magic answer to this most basic question. After surveying CSAs all over the country, though, Tim Laird came up with some practical recommendations. First, before you even begin growing the food, discuss with your members how much they want. Spend at least a winter planning. Next, do not overload the shares. More is not necessarily better. Especially when beginning, you need to *plant* extra to be sure to have enough, but you do not have to *force* all your food on your sharers. Excessive amounts of produce lead sharers to split shares or to drop out in embarrassment over the waste composting in their refrigerators. Finally, organize to offer some choices, if possible. This is easier when sharers pick up at the farm or at a farm stand. (See chapter 15 for ideas on how CSAs arrange for choices and chapter 14 for a discussion of quality.)

Getting the right quantity of food can take some adjustment. Robyn described her experience at Indian Line Farm: "We started out intending a share to feed two to four people with a mixed diet or one

Seasonal Highlights:

The Intervale Community Farm grows over 35 different vegetables, 20 different flowers, a dozen types of herbs, berries, and melons. We harvest from early June through mid-November. Each week you'll receive a diverse mix of those crops that are in season. While no two years are the same, here are some of the typical highlights.

June: Luscious lettuce, succulent salad greens, plentiful peas, and bountiful broccoli. This is salad month, a fresh start to the harvest year.

July: Spinach and greens are replaced by baby beets, crisp carrots, and crunchy cukes. Mid-month arrives with a selection of summer squash, green beans, and basil. By late July we'll have onions, and if we're lucky, tomatoes.

August: The dog days...peppers, eggplant, corn, onions, and new potatoes all arrive. Tomato sandwiches grace your picnics, and by mid-month: melons. Bikers find it hard to carry everything home!

Sept: Fall raspberries are a special treat. Cool nights invite the return of greens and boost broccoli, while warm days keep the August crops happy. Vegetable medleys and simple steamed veggies adorn your table.

October: Frost ends the summer vegetables, while improving the flavor of those remaining. Crunchy carrots and cauliflower, bowls of potato-leek soup, stuffed and steaming winter squash. Pumpkins for carving and eating.

Nov: The season ends with cold-loving kale, greens, carrots, beets, potatoes, and cabbage. Hearty stews and roasted root vegetable collages fend off the impending winter.

This page from the brochure of the Intervale CSA in Burlington, Vermont, shows the seasonal variety of shares.

to two vegetarians. We determined the 10 to 15 pound weekly average from the total weight we figured the household would need over the entire year. This turned out to be way too much for most households, especially as we distributed the shares twice a week. The next year, people found somebody to take the other half, so in the third year we cut back to one distribution a week. Even at the reduced size, some people wanted a smaller portion. So we provided half-shares at a little more than half the price."

Farmers, who are used to their own capacious appetites, sometimes have a hard time believing city people can eat so little. The best way to establish the share size is to learn from your sharers how much they want. Take an annual poll of the members. Boise Food Connection and Wishing Stone Farm have good membership questionnaires. In the GVOCSA vegetable questionnaire, we explain that members will be receiving seven to eight items a week. Given that, we ask how much they would prefer to receive of each vegetable or fruit. Surprisingly, most of them want similar quantities of most items. Bunched greens and root crops, such as turnips and rutabagas, arouse the widest deviations. Some CSAs provide as many as fifteen or twenty items a week.

Michael Docter of the Food Bank Farm in Massachusetts claims that the reluctant greens eaters among his sharers came to view his well-meaning attempts at increasing their greens intake as a "colossal collard conspiracy." The average between "none" and "big bunches" may be a small bunch, which won't really satisfy either group, so the GVOCSA found an alternative solution. Members who find that their shares have too many greens or roots can put them in the "Share box" for others to take. Those who get too few can order extra through the weekly bulk order. Magic Beanstalk CSA offers a greens share of an extra bunch a week. CSAs that have pickup at the farm can solve this more simply by allowing sharers to take what they want up to a given limit—the "Mix-and-Match" system.

What do YOU like?

Please indicate your preferences and dislikes to help us adjust our planting.

1. Most favored vegetable status
2. A favorite, use large quantities
3. Use moderate amounts
4. A little goes a long way
5. I wouldn't eat it when I was a kid, and I won't eat it now.

Mesclun	1	2	3	4	Sweet Corn	1	2	3	4
Lettuces	1	2	3	4	Tomatoes	1	2	3	4
Radishes	1	2	3	4	*Low acid*	1	2	3	4
Salsify	1	2	3	4	*Cherry*	1	2	3	4
Kale	1	2	3	4	*Paste, vine*	1	2	3	4
Spinach	1	2	3	4	Zucchini	1	2	3	4
Baby Beets	1	2	3	4	Summer Squash	1	2	3	4
Baby Carrots	1	2	3	4	Acorn Squash	1	2	3	4
Broccoli	1	2	3	4	Butternut Squash	1	2	3	4
Diakon Radish	1	2	3	4	Cabbage	1	2	3	4
Pole Beans	1	2	3	4	Tatsoi	1	2	3	4
Green Beans	1	2	3	4	Leeks	1	2	3	4
Basil	1	2	3	4	Onions	1	2	3	4
Bell Peppers	1	2	3	4	Asparagus	1	2	3	4
Hot Peppers	1	2	3	4	Potatoes	1	2	3	4
Cucumbers	1	2	3	4	Pumpkins	1	2	3	4
Eggplant	1	2	3	4	Turnips	1	2	3	4
Shell Peas	1	2	3	4	Radicchio	1	2	3	4
Snap Beans	1	2	3	4	Winter Carrots	1	2	3	4
Watercress	1	2	3	4	Fennel Bulbs	1	2	3	4

Vegetable preference questionaires, like this one from Wishing Stone Farm in Rhode Island, help CSAs plan.

To get the highest level of member response, Angelic Organics includes a postcard in their share boxes every week, giving members the chance to comment immediately on quality and quantity. Of course, farmers who see their sharers face to face on a regular basis can get this information directly.

Boise Food Connection — ***Subscriber Preferences Survey***

Please fill out this survey and return it to the address on the reverse side. The information you provide will help us develop subscription offerings that best meet your needs.

1 ☐ **Yes, please send me a subscriber prospectus for Spring/Fall 1993.**

Name & Address (please print)
2
3
4
5

Telephone number
6

7 **Number of people in your household**

Number of meals eaten at home per week
8 **Breakfast**
9 **Lunch**
10 **Dinner**

Would you be willing to help out in any of the following areas?

11 ☐ Planning Committee
12 ☐ Harvest
13 ☐ Sorting
14 ☐ Pickup & Delivery
15 ☐ Newsletter / Fact Sheets
16 ☐ Other:

Are you interested in doing any of the following to preserve your seasonal foods?

17 ☐ Canning
18 ☐ Freezing
19 ☐ Dehydrating (drying)

Please indicate which of the following items you'd be interested in purchasing on a subscription basis

Vegetables
20 ☐ Asparagus
21 ☐ Beans
22 ☐ Beets
23 ☐ Broccoli
24 ☐ Cabbage, red
25 ☐ Cabbage, green
26 ☐ Carrots
27 ☐ Cauliflower
28 ☐ Celery
29 ☐ Celeriac
30 ☐ Chard
31 ☐ Chinese Cabbage
32 ☐ Corn, sweet
33 ☐ Collards
34 ☐ Cucumbers
35 ☐ Eggplant
36 ☐ Garlic
37 ☐ Kale
38 ☐ Leeks
39 ☐ Lettuce
40 ☐ Mustard
41 ☐ Onions, green
42 ☐ Onions, fresh sweet
43 ☐ Onions, storage
44 ☐ Okra
45 ☐ Parsnips
46 ☐ Peas
47 ☐ Snow Peas

Vegetables
48 ☐ Snap Peas
49 ☐ Peppers
50 ☐ Potatoes, new
51 ☐ Potatoes
52 ☐ Pumpkins
53 ☐ Radish
54 ☐ Rutabaga
55 ☐ Scallions
56 ☐ Spinach
57 ☐ Squash, summer
58 ☐ Squash, winter
59 ☐ Sweet Potatoes
60 ☐ Tomatoes
61 ☐ Turnips

Fruits
62 ☐ Cantelope
63 ☐ Watermelon
64 ☐ Rhubarb
65 ☐ Strawberries
66 ☐ Raspberries
67 ☐ Huckleberries
68 ☐ Blackberries
69 ☐ Grapes
70 ☐ Apple
71 ☐ Apricot
72 ☐ Cherry, sweet
73 ☐ Cherry, pie

Fruits
74 ☐ Nectarine
75 ☐ Peach
76 ☐ Pear
77 ☐ Plum

Herbs
78 ☐ Arugula
79 ☐ Basil
80 ☐ Chives
81 ☐ Cilantro
82 ☐ Dill
83 ☐ Marjoram
84 ☐ Oregano
85 ☐ Parseley
86 ☐ Rosemary
87 ☐ Sage
88 ☐ Thyme

Grains
89 ☐ Dry Beans
90 ☐ Wheat
91 ☐ Barley
92 ☐ Oats
93 ☐ Corn
94 ☐ Popcorn

Nuts
95 ☐ Walnuts

Meats
96 ☐ Chicken
97 ☐ Turkey
98 ☐ Beef
99 ☐ Lamb
100 ☐ Pork
101 ☐ Fish

Dairy
102 ☐ Eggs
103 ☐ Milk
104 ☐ Cheese
105 ☐ Yogurt

Misc
106 ☐ Flowers
107 ☐ Honey

Prepared Items
108 ☐ Wine
109 ☐ Cider
110 ☐ Apple Juice
111 ☐ Jam
112 ☐ Jelly
113 ☐ Flour, fresh milled
114 ☐ Sausage

115 **Other Items**

Ver 1.00 (1/9/93)

Source: Boise Food Connection.

David Trumble at Good Earth Farm in Weare, New Hampshire, offers his local members a novel choice: They can sign up for either the "Dave's Mix" or "Member Choice" option in either family (full-share) or single-sized (half-share) boxes. Those with Member Choice pick what they want from a computer-generated list in the previous week's box, up to a certain number of units, and call the farm by the cutoff time to place their order. The farm keeps a second copy of the lists by the phone: Dave or his wife Linda check off the orders as members call them in. If members forget, they simply receive a "Dave's Mix" box for that week. This system also gives the farm a better grasp of how often to put vegetables in all the shares. Close to half of the local sharers take the Member Choice option, a total of about twenty a year, which Linda reports is manageable. (Trying to do this for over one hundred shares drove Cass Peterson and Ward Sinclair to give up running a CSA.) The Member Choice costs $80 extra for the single size and $100 extra for the family size. Ben Watson, the editor for this book, belongs to Good Earth Farm and, when he did not see this system mentioned in the draft version, wrote:

> When I first joined Dave's CSA back in 1992, he didn't have the option of making your own selection, and my then-girlfriend and I didn't know what to make of Swiss chard week in and week out. I find the Member Choice option works better for me, particularly since I have my own garden, and thus don't need many peas, beans, tomatoes, and other crops. Instead, I supplement my own garden's production with Dave's (he grows better salad greens, potatoes, and carrots than I do, plus space-hogging crops like winter squash). In fact, I dropped out of his CSA in 1993, and then re-upped when he instituted the Member Choice and single-share size. Without these two options, I would not have remained a member, despite my commitment to local agriculture.

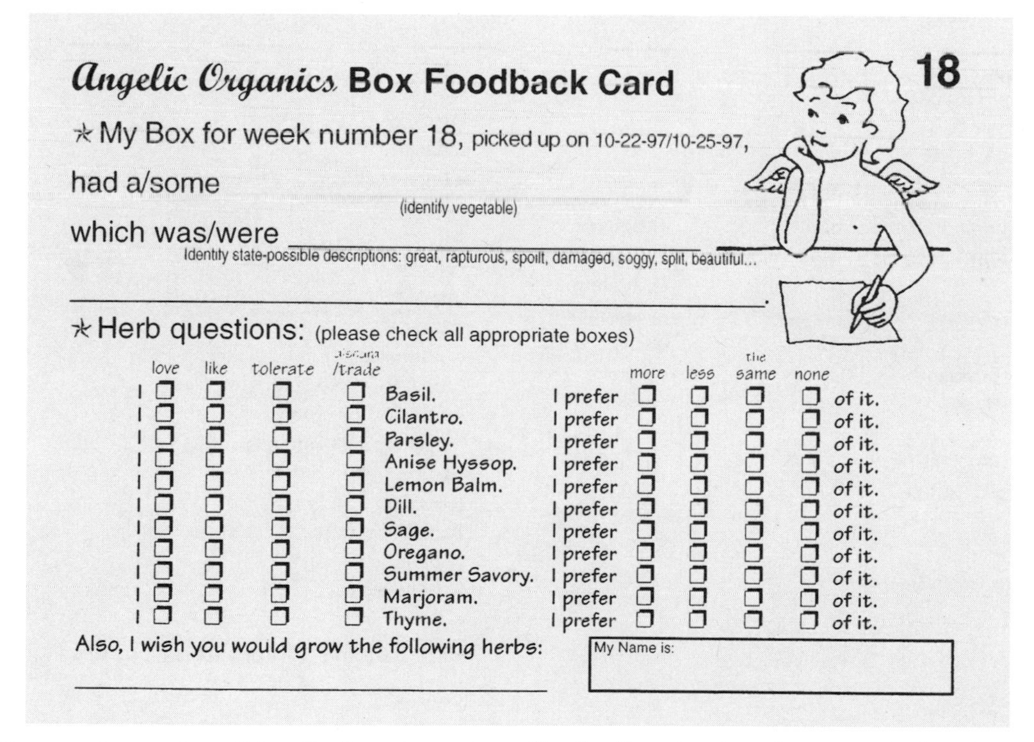

Angelic Organics Box Foodback Card 18

★ My Box for week number 18, picked up on 10-22-97/10-25-97,

had a/some ____________________
(identify vegetable)

which was/were ____________________
Identity state-possible descriptions: great, rapturous, spoilt, damaged, soggy, split, beautiful...

____________________.

★ Herb questions: (please check all appropriate boxes)

	love	like	tolerate	/trade			more	less	the same	none	
I	☐	☐	☐	☐	Basil.	I prefer	☐	☐	☐	☐	of it.
I	☐	☐	☐	☐	Cilantro.	I prefer	☐	☐	☐	☐	of it.
I	☐	☐	☐	☐	Parsley.	I prefer	☐	☐	☐	☐	of it.
I	☐	☐	☐	☐	Anise Hyssop.	I prefer	☐	☐	☐	☐	of it.
I	☐	☐	☐	☐	Lemon Balm.	I prefer	☐	☐	☐	☐	of it.
I	☐	☐	☐	☐	Dill.	I prefer	☐	☐	☐	☐	of it.
I	☐	☐	☐	☐	Sage.	I prefer	☐	☐	☐	☐	of it.
I	☐	☐	☐	☐	Oregano.	I prefer	☐	☐	☐	☐	of it.
I	☐	☐	☐	☐	Summer Savory.	I prefer	☐	☐	☐	☐	of it.
I	☐	☐	☐	☐	Marjoram.	I prefer	☐	☐	☐	☐	of it.
I	☐	☐	☐	☐	Thyme.	I prefer	☐	☐	☐	☐	of it.

Also, I wish you would grow the following herbs: ____________________

My Name is:

Every week, members of Angelic Organics receive a "foodback" card such as this one with their share.

WEEKLY SHARES THROUGH THE SEASONS

Spring Crops

Vegetables	*Herbs*	*Fruit*
asparagus	oregano	strawberries
dandelion greens	dill	
mizuna	cilantro	
arugula	tarragon	
radishes	chives	
garlic greens	garlic chives	
spinach	thyme	
corn salad (mâché)		
turnip greens		
bok choi		
lettuce		
mustard greens		
radicchio		
Chinese cabbage		
tatsoi		
broccoli raab		
Hon tsai tai		
senposai		
cress		
snow peas		
sugar snaps		
beets—root and greens		
scallions		
shelling peas		
escarole		

Summer Crops

Vegetables	*Herbs*	*Fruit*
lettuce	dill	blueberries
Swiss chard	basil	raspberries
green onions	summer savory	gooseberries
carrots	marjoram	cantaloupe
kale		watermelon
purslane		blackberries
vegetable amaranth		cherries
summer squash		peaches
cucumbers		plums
beans—green, purple, Romano, yellow		nectarines
		apricots
sorrel		
chicory		
garlic tops (scapes)		
peppers		
eggplant		
tomatoes		
early potatoes		
cabbage		

Summer Crops *(continued)*

Vegetables	*Herbs*	*Fruit*
okra		
sweet corn		
tomatillos		
broccoli		
cauliflower		
lima beans		
butter beans (green soy beans)		
fava beans		

Fall Crops

Vegetables	*Herbs*	*Fruit*
broccoli	garlic	apples
kohlrabi	parsley	pears
onions	cilantro	
shallots	dill	
potatoes	sage	
celery		
daikon radishes		
leeks		
carrots		
beets		
spinach		
chickweed		
bok choi		
tatsoi		
radicchio		
Chinese cabbage		
pumpkins		
winter squash		
collards		
mizuna		
arugula		
cauliflower		
cabbage—green, red, savoy		
burdock root (gobo)		
long-keeping tomatoes		
lettuce		
Jerusalem artichokes		
Brussels sprouts		
celeriac		
scorzonera/salsify		

Late Fall and Winter Storage Crops

Vegetables	*Herbs*	*Fruit*
broccoli	garlic	apples
potatoes	parsley	pears

Vegetables	*Herbs*	*Fruit*
Late Fall and Winter Storage Crops *(continued)*		
onions	cilantro	
carrots		
leeks		
turnips		
rutabagas		
Brussels sprouts		
kale		
collards		
spinach		
parsnips		
celeriac		
winter squash		
cabbage		
Chinese cabbage		
Hispanic Shares		
carrots	garlic	[pineapple]
lettuce	wild mustard with seed	[bananas]
onions	purslane	watermelon
potatoes	sweet corn	cantaloupe
cucumbers	cilantro	cherries
tomatillos	radishes	apples
tomatoes (Roma)	spinach	peaches
beets	avocado	apricots
broccoli	jicama	mango
cauliflower		
summer squash & squash flowers		
green beans		
cabbage		
peppers (sweet green, frying, and hot)		
nopales		

Macrobiotic Shares

macrobiotic diets usually do not *include:*
tomatoes
potatoes
eggplant
peppers
spinach
Swiss chard
some also exclude:
corn
zucchini
lettuce, except for romaine

On the other hand, Bette Lacina, who farms on 1 1/2 acres at Under the Willow Produce in Sag Harbor on Long Island, presents a compelling argument *against* choice:

> Our concept was to tell people we are experimenting with growing on a small piece of land in an urban area where land is extremely expensive. If you want to support what little open land is left, you need to be willing to take what we can produce and what nature gives us. If it's a dry season or a wet season, we'll have different things. We can't guarantee anything. And the response I've gotten has been tremendous. I'm learning to eat things I never ate before. A lot of people say to me at the end of the season, they *like* not having a choice. It's a challenge to creative people.

To help you design your shares, we are including a list of vegetables, herbs, and fruit by season, which includes some fairly unusual vegetables along with the old garden standards. To keep CSAers in the North from turning green with envy, we have left out exotics such as figs, persimmons, avocados, and artichokes. If you can grow these, by all means put them in your shares. Farms with nearby Hispanic communities might try designing special shares. We have also listed choices for a macrobiotic share: by general agreement, solanaceous vegetables (potatoes, tomatoes, eggplant, peppers) are out, while controversy continues over such items as corn, zucchini, and lettuce. Check with your members, if they are interested in this option. In addition, we have chosen a few sample share descriptions from Winter Green Farm in Noti, Oregon, Caretaker Farm in Williamstown, Massachusetts, and Mill Valley in Stratham, New Hampshire.

Let your imagination devise creative forms of shares. One little CSA grows ingredients for spaghetti sauce. The members gather in late summer to cook and can their sauce together. Many farms offer flower shares, a certain number of bouquets for a set price paid in advance. Separating shares by season seems to work for some farms that provide summer and

Share Information

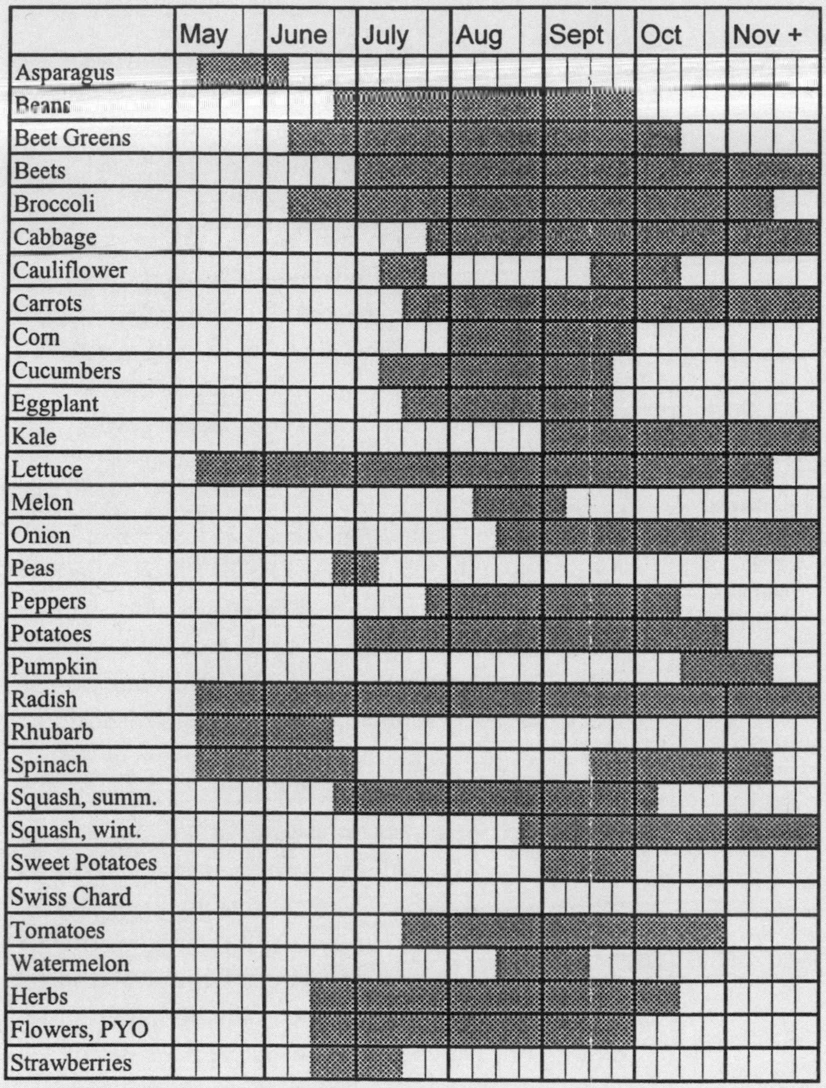

	May	June	July	Aug	Sept	Oct	Nov +
Asparagus							
Beans							
Beet Greens							
Beets							
Broccoli							
Cabbage							
Cauliflower							
Carrots							
Corn							
Cucumbers							
Eggplant							
Kale							
Lettuce							
Melon							
Onion							
Peas							
Peppers							
Potatoes							
Pumpkin							
Radish							
Rhubarb							
Spinach							
Squash, summ.							
Squash, wint.							
Sweet Potatoes							
Swiss Chard							
Tomatoes							
Watermelon							
Herbs							
Flowers, PYO							
Strawberries							

Crop	Amt/ share	Units
Basil	0.56	lbs
Beans, green	9.76	lbs
Beans, purple	1.88	lbs
Beans, yellow	3.21	lbs
Beet Greens	0.97	lbs
Beets	14.6	lbs
Broccoli	4.88	lbs
Cabbage	3.83	lbs
Carrot, 2nd	1.24	lbs
Carrots	34.2	lbs
Celery	5.09	lbs
Chard	1.43	lbs
Corn	2.79	lbs
Eggplant	4.3	lbs
Kale	1.47	lbs
Leeks	0.41	lbs
Lettuce	33.9	heads
Melons	19.2	lbs
Mesclun greens	1.45	lbs
Onions	12.8	lbs
Parsley	0.24	lbs
Peas	4.9	lbs
Peppers, grn	5	lbs
Peppers, hot	0.82	lbs
Pickling cukes	6.47	lbs
Potatoes	34.2	lbs
Radish	2.28	lbs
Rhubarb	7.72	lbs
Summer squash	32.7	lbs
Slicing cukes	8.66	lbs
Snap peas	1.92	lbs
Spinach	2.9	lbs
Strawberries	3.86	qts
Tomatoes	36	lbs
Tomatoes, cherry	2.12	lbs
Tomatoes, green	0.66	lbs
Tomatoes, plum	4.43	lbs
Watermelon	14.3	lbs
Winter squash	15.9	lbs

Source: Mill Valley Farm CSA.

winter shares. Tuscaloosa CSA divides its produce into spring/fall shares and summer shares. Seven Springs Farm in Virginia grows the basic share and contracts with other farms for a few crops, such as sweet corn, and for "sub-shares." In addition to the regular share, members can purchase a "gourmet sampler," twenty-five pounds of French filet beans, baby vegetables, shiitake mushrooms, and the like. In Portland, Oregon, John Martinson and Beverly Koch offer "The Birds and Bees Option," a mouth-watering array of seventeen fruits and berries, nuts, honey, and eggs, as an add-on to Urban Bounty CSA shares. A CSA called Own-a-Goat in Virginia allows goat producers to distribute raw goat's milk to people who want to buy it. The law prohibits raw milk sale, but not drinking milk from a goat you "own," even though a farmer cares for it and milks it (*FARM* [Summer/Fall 1996]: 1). Silver Creek Farm in Hiram, Ohio, offers "Preserver shares," and received SARE funding to set up a canning shed at the farm and to provide lessons in canning. They also sell beer shares for homebrewers which include hops, malt, sugar, and yeast. Each year a different brewmaster sets the recipe: everyone who signs up shares in the cost and receives three six-packs of beer. (For ideas on connections with other farms, see chapter 18, "Regional Networking for Farm Products.")

CSAs can offer meat, eggs, and milk as part of the regular share, or create separate shares for vegetarians and meat eaters. Several farms can associate to supply a wider array of products. Silver Creek Farm sells shares in its own lamb, chicken, and eggs, but also goat's-milk cheese and beef shares from neighboring farms. Rather than recruiting more members, CSAs can grow by diversifying product offerings to provide more of the food needs of current members. Wayback Farm in Belmont, Maine, has taken this path, gradually adding milk, meat, dry beans, and grains to its vegetable shares.

Sharers' abilities to handle the flow of food vary greatly. By supplying recipes and tips on storage, the CSA newsletter can help members improve their skills. The GVOCSA newsletter has featured stories on members who have systems for blanching and freezing extra greens or have found ways to turn their porch into a root cellar. The days are gone when everyone grew up snapping beans and stocking the family root cellar. Broadcasting from Muncie,

COMMUNITY FARM PROJECTED CROPS PER LARGE SHARE

1995 GROWING SEASON			JUNE			JULY				AUGUST					SEPT				OCT			
CROP	TOTAL	WEEKLY	12	19	26	4	11	18	25	1	8	15	22	29	5	12	19	26	3	10	17	24
basil	1 lb+							❧	❧	❧	❧	❧	❧	❧								
beets	4 lbs	1 lb			❧		❧													❧		❧
blueberries	16 pts	2 pts						❧	❧	❧	❧	❧	❧	❧	❧							
broccoli	20 lbs	1 lb	❧	❧	❧	❧	❧	❧	❧	❧	❧	❧	❧	❧	❧	❧	❧	❧	❧	❧	❧	❧
burdock	1.5 lbs	.5 lb															❧		❧		❧	
cabbage	4 heads	1 head			❧												❧		❧		❧	
carrots	30 lbs	1.75 lbs				❧	❧	❧	❧	❧	❧	❧	❧	❧	❧	❧	❧	❧	❧	❧	❧	❧
cauliflower	7 heads	1 head				❧		❧		❧		❧		❧		❧		❧				
chinese cabbage	3 heads	1 head				❧													❧			❧
corn	48 ears	8 ears											❧	❧	❧	❧	❧	❧				
cucumbers	30 fruit	3 fruit							❧	❧	❧	❧	❧	❧	❧	❧	❧	❧				
eggplant	8 fruit	1 fruit										❧	❧	❧	❧	❧	❧	❧	❧			
endive	3 heads	1 head		❧														❧		❧		
flowers	occasionally																					
french sorrel	6 bun.	1 bun.	❧		❧		❧											❧		❧		❧
garlic	40 bulbs	40 bulbs											❧									
green beans	4 lbs	2 lbs						❧	❧													
green onions	8 bun.	1 bun.	❧	❧	❧	❧	❧	❧	❧	❧												
kale	5 bun.	1 bun.																❧	❧	❧	❧	❧
leeks	15 stalks	3 stalks																❧	❧	❧	❧	❧
lettuce	40 heads	2 heads	❧	❧	❧	❧	❧	❧	❧	❧	❧	❧	❧	❧	❧	❧	❧	❧	❧	❧	❧	❧
melons	15 fruit	3 fruit												❧	❧	❧	❧	❧				
red onions	6 lbs	1 lb									❧	❧	❧	❧	❧	❧						
storage onions	18 lbs	9 lbs												❧								❧
pak choi	2 heads	1 head																	❧		❧	
parsley	11 bun.	1 bun.	❧		❧		❧		❧		❧		❧		❧		❧		❧		❧	❧
peas	2.5 lbs	1.25 lbs	❧		❧																	
new potatoes	20 lbs	10 lbs				❧		❧														
storage potatoes	40 lbs	20 lbs													❧							❧
pie pumpkins	3 fruit	3 fruit																				❧
jack-o-lanterns	2 fruit	2 fruit																				❧
spinach	8 bun.	1 bun.	❧		❧		❧		❧						❧		❧		❧		❧	
summer squash	25 fruit	2.5 fruit							❧	❧	❧	❧	❧	❧	❧	❧	❧	❧				
sweet peppers	24 fruit	2.5 fruit							❧	❧	❧	❧	❧	❧	❧	❧	❧	❧				
tomatoes	50 lbs	4 lbs						❧	❧	❧	❧	❧	❧	❧	❧	❧	❧	❧	❧			
winter squash	20 fruit	2/2wks+16																		❧		❧

Source: Winter Green Community Farm.

Indiana, "A Prairie Home Companion's" Garrison Keillor once referred reverentially to the town as the "Mother Church" of canning, since it was the home of Ball canning jars. With his inimitable logic, Keillor blamed the failure of modern parents to can their own tomatoes as the cause of the destruction of their children's education. According to Keillor, in the days when parents canned, they were able to afford to send their children to Ivy League colleges.

Modern homes and apartments have very limited food storage space. Many CSAs have tried to organize "processing parties" to teach the basics of canning, freezing, or dehydrating, but most have not been successful; members are simply too busy to attend workshops. In some states, the Extension Service offers workshops and advice on food storage skills. The CSA newsletter could carry a schedule of such local offerings.

After a few years of answering members' questions about odd vegetables, my partner David and I wrote *FoodBook for a Sustainable Harvest*. Lovely illustrations by Karen Kerney enabled readers to identify mystery crops. Entries for seventy crops include history, nutritional information, short- and long-term storage requirements, anecdotes from growing at Rose Valley Farm, and recipes, many contrib-

THE PRODUCTION OF FOOD IN THE SERVICE OF LAND AND COMMUNITY

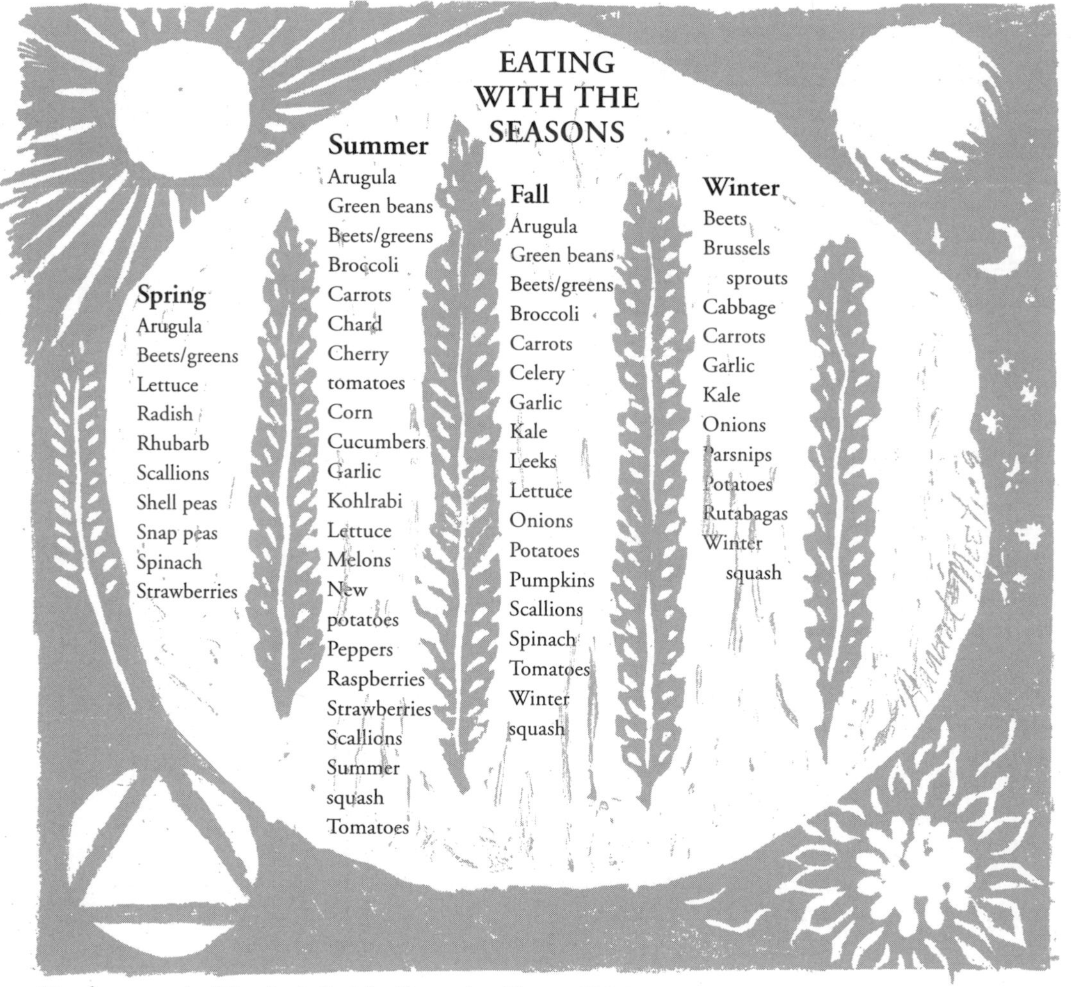

Linoleum cut by Elizabeth Smith, Caretaker Farm, 1997.

ELIZABETH HENDERSON

On a rainy morning, Genesee Valley Organic CSA sharers pack boxes of vegetables and flowers into their cars to transport them back to town for distribution.

uted by members. Inspired by our example, the Madison Area Community Supported Agriculture Coalition (MACSAC) created *From Asparagus to Zucchini: A Guide to Farm-Fresh, Seasonal Produce,* beautifully illustrated by Bill Redinger. It contains essays on eating locally, write-ups with recipes for forty-six vegetables, profiles of the MACSAC member farms, and other resources. (You will find information on how to acquire these books in CSA Resources.)

In "Eating Seasonally: A Personal Reflection," David Bruce paints with bold strokes the broad cultural, historical, anthropological, and cosmic significance of this practice:

> Those who are stepping into seasonal eating by participating in a CSA deserve recognition for their extra efforts. It is revolutionary eating, refusing to be a part of the environmental degradation that characterizes the current food system. It is charitable eating, wanting those who produce your food to be earning a decent standard of living, and so loving others as yourself. It is Zen eating, requiring mindfulness, a simplifying of and a concentration on what we consume. It is environmental eating, for what better action than to provide an example to those around you on living more harmoniously in your world? With revolutionary spirit, charitable hearts, thoughtful practice, and sensitive action, we can learn to eat seasonally. As we do so, we learn about our region and, with creativity and caring responsibility to place, we can develop our own regional epicurean faire. (*From Asparagus to Zucchini*, pp. 14–15)

The most fitting conclusion for this chapter on the food is "An Ode to Community Supported Agriculture," by Anna Barker:

Beauty and the Feast

In December air
Indoors growing together
Snow mist frosts the fields

We are cooking.
Recipes for diversity
must include hair-brained schemes and land purchases,
awareness of trends, open houses with salsa (home-canned,
of course), sizzle, regional/seasonal ingredients, and Trust.

New Pioneers in the MERF Movement (and elsewhere, where ever*
there is Common Ground) who receive grants and give advice
along with healthy food, know about efficiency, keep their knives
sharp, and enjoy the flavor of the batter in their stories.

We "walk the talk," reach out to network and touch 1-800-516-7797
while "cultivating education mechanisms with consumer." Really.

Trust, grown from self-delivered knowledge (as in midwifery) and
experience-based education result in believing and form the basis
for growing the caring/carrying capacity. (Money counts.)

See us: we are not a fad or a social echo; we are The Land.
We are a community of sustainers, co-creators of a Way of Life,
setting loose a system of nutrition based on the soil, nurtured
by the craft of farming, by elders grounded in integrity, giving
witness through words meeting actions.

Join us as we regain the heritage that lies deep in the Earth,
our Mother, much-covered but alive beneath layers of concrete,
asphalt, and square buildings.

Chew works, spit out ideas. Let the organic process spread,
dynamically influencing source-labels that increase Trust.
Through facts and knowledge gained from experience
stimulate growth-through-the-seasons;
grow as individuals, as unique as your foot prints on soil after rain.
Participate in peeling back the "price per pound" mentality and
share in linking lives with The Land in ways that cultivate
community along with food security.

Eat close to home, and invite many guests to share the feast.

In the cool spring air
Outdoors growing together
December warms fields

This ode (framed by two haiku), is a synthesis of words chosen by participants in "Eating Close to Home," the Upper Midwest CSA Conference held December 2–4, 1994, at the University of Wisconsin-River Falls, to summarize their experience in Saturday's workshop. Anna Barker, conference attendee and poet, created the synthesis and haiku. Reprinted with permission from the Minnesota Food Association *Digest* (December 1994): 2.

* MERF - Madison Eaters Revolutionary Front

COMBINING CSA WITH OTHER MARKETS 17

The goal must be to raise all food in wholesome, sustainable ways while eliminating the poverty and inequality that deprive many of the ability to buy an adequate diet at any price. Food produced on prosperous and sustainable family farms should be the affordable food of choice for ordinary people everywhere.

—MARTY STRANGE, "PEACE WITH THE LAND, JUSTICE AMONG OURSELVES"

ALTHOUGH IDEALLY CSA FARMS might be totally devoted to their members, in reality most farms and gardens must sell to other markets as well. Some farms retain other established markets when initiating a CSA. Others, begun as dedicated CSAs, find that they need supplemental income and have excess production that their sharers cannot comfortably absorb. Too much produce in the weekly share has driven off more CSA members than too little.

Tim Laird's survey in 1993 found that 74 percent of the CSA farms were selling to outside markets. Some farmers felt that maintaining diverse markets increased their security; others planned to give up outside sales as soon as they found enough sharers to ensure financial survival. Outside markets included restaurants, farmers' markets, food co-ops, local stores, and natural foods supermarkets. I learned at the NOFA Vermont conference in 1998 that only one of the thirty CSAs in Vermont produces exclusively for members. The statistics in California are similar, where CSAs supplement other marketing outlets. With close to six hundred shares making up one-third of the sales of Full Belly Farm, Dru Rivers says it may be a "watered-down" version of CSA, but she feels very good about it. Through careful record keeping, Richard de Wilde at Harmony Valley has calculated that the specialty crops, such as celery root and greens, which they grow for distributors, net more income than their 335 shares. He says the CSA is valuable for reasons other than the money.

Since the early 1980s, food co-ops in many parts of the country have been the most faithful markets for local organic farms. Some, like the Wedge Co-op in Minneapolis, Minnesota, make preseason contracts with growers for future crops. At winter meetings, the co-op and the growers review the previous season and agree on crop projections. They also set prices for the year to come at a rate produce manager Ed Brown calls the "Grower Sustainable Price," a premium that accounts for the higher cost of producing food in Minnesota. Brown wants to buy as much as possible from local farmers and is willing to educate his customers about local farming conditions. He also lines up second-string growers from whom he buys if the first string has production difficulties, so the co-op does not share the risk with the farmers as many CSAs do.

In 1993, 74 percent of the CSA farms surveyed were selling to outside markets. Some farmers felt that maintaining diverse markets increased their security; others planned to give up outside sales as soon as they found enough sharers to ensure financial survival.

The CSA where Tim Laird worked, Quail Hill Community Farm, markets to some restaurants. One restaurant purchases bulk quantities of a few crops, while another takes two shares and features Quail Hill produce on its menu. Following the lead of Alice Waters of Chez Panisse fame, chefs in major cities around the country have increased their purchases of fresh, in-season produce from local farms. A few restaurants and schools for chefs, such as the New England Culinary Institute in Burlington, Vermont, and Common Ground, a cooperative restaurant in Brattleboro, Vermont, purchase regularly from local farmers and meet with them every winter to discuss specific crops and quantities (see Beth Holtzman's "The Intervale" in chapter 21 for more details). This incipient Restaurant Supported Agriculture (RSA) has yet to blossom into widespread advance contracting and payments to farms.

This may change if the Catskill Watershed Farms RSA takes off as planned. In the spring of 1998, Hilary Baum and Rick Bishop, of the New York City Watershed Agricultural Council, invited chefs and specialty retailers such as Balducci's to join a partnership with watershed farms. They asked the chefs to make a commitment to preseason planning of crops, quantities, and prices, to purchase the products if specifications were met, to communicate regularly with the project, and to pay the farms in a timely way. On their side, the Watershed Council promised to certify the producers as participants in "Whole Farm Planning" to prevent pollution of New York City's drinking water, to help farms meet the preseason plans, to deliver crops in peak condition, to keep restaurants informed of any deviation from the delivery schedule, to offer product "specials" and media opportunities, and to provide a newsletter on how production and the program is progressing. Underlined in bold print, the proposal states that it is not asking buyers to pay in advance, and concludes: "Support our effort to preserve family farming in our region, while we protect our drinking water and bring you quality food and good service. *You'll be doing a great thing!*"

How do you decide what produce goes to your sharers and what to sell to other markets? Farms accustomed to selling to outside markets can easily calculate this by setting a dollar limit on the weekly share. For Rose Valley in 1997, that limit was $12.50 a week based on our wholesale prices. Each week, I planned a share that consisted of at least seven or eight items that would have sold at the farmers' market for $12.50. If the share was short one week, I made up the difference another week. The consumer-run Marblehead Eco-Farm CSA, in Massachusetts, aims to pay the farmers who supply their produce 80 percent of the highest retail price they can find for an item in local stores. If you are not familiar with market prices, a set number of items and a set number of pounds could be your guide. Any produce that remains after the shares are filled is available for sale elsewhere.

When we began the GVOCSA, the 29 shares consumed a small percentage of our production at Rose

Valley. Gradually, the number of shares grew to 164, which accounted for close to 85 percent of that year's production. We did not devote a particular section of our ground to the CSA. Instead, we divided production into crops grown exclusively for the CSA and crops for both the CSA and outside markets. The Veggie Questionnaire establishes the amount of these crops members want per week. In periods of great abundance, we give them up to one and a half times the optimum share. For example, most of our sharers agree that 1 pound of beans per week is a good supply, so we never give them more than 1½ pounds of beans at a time. Any beans beyond that go to other markets. (See chapter 13 for the quantities several CSAs plan to supply.)

In his survey, Tim Laird found growers who make a much clearer distinction between CSA and market crops. One farm had "separate crews do the planning and work, very independent of each other." Other farms designated particular crops, such as white eggplants, yellow tomatoes, garlic, or fall crops, for outside markets. Martha Jacobs at Slack Hollow Farm in Argyle, New York, who runs a CSA of thirty shares along with her main production for other markets, found that trying to grow a few rows of CSA crops along with her wholesale crops led to confusion. Trying to save a row of kale amidst many beds of harvested lettuce can lead to a weedy mess. So Martha sets off one acre for CSA crops and treats it like a big garden, with many crops grown on a small scale rotating within that acre.

In addition to running their own CSA, Slack Hollow grows the carrots and onions for the Hudson-Mohawk CSA, an important connection for both

ELIZABETH HENDERSON

Gleaners at Rose Valley Farm.

mutual support and friendship. Janet Britt and her crew enjoy helping with planting and harvesting, and pay Slack Hollow for the onions in advance. The year when Slack Hollow lost all their onions to disease, the two CSAs suffered together, though Slack Hollow was able to soften the loss by substituting leeks and garlic. The moral Martha draws is the importance of being flexible. Slack Hollow is expanding carrot production to supply two other CSAs as well.

The real crunch can come during times of shortage. Drought, flood, disease—any of the plagues the Sovereign of the Universe inflicted upon the Egyptians—can descend upon a farm. Do you give as much as possible to your sharers, or do you supply the other markets for money that the farm may need badly? For the GVOCSA, the CSA always comes first. GVOCSA members sign a contract for the entire season, and pay in advance of receiving our produce. None of our other markets gives us this level of commitment. Many other CSA farmers feel the same way. Tom Meyers, who runs the market garden at Hawthorne Valley, observes that if you short other markets, you feel it in the pocketbook immediately; if you short the CSA, you'll feel it next year, when the membership dwindles.

Winter Green Community Farm Noti, Oregon

In 1992, the year of the big rains, we cut out all other markets, bought some extra boots, and members slogged with us through the mud. Share packets were heavy on survivor greens like kale and collards, but no one complained. The warm and steady support from our members was wonderful for our morale and helped us through what could have been a devastating time.

Quality problems can also give you pause in deciding where or whether to sell a crop. Good communications with your CSA members may enable you to give them vegetables that you cannot sell elsewhere for aesthetic reasons. Late sweet corn that was delicious, but wormy, threw Martha Daughdrill, who runs Newburg Vegetable Farm with her husband in Maryland, into a quandary:

> It tasted so fantastic, but it was wormy up the kazoo! We took a little bit to market, but we sold hardly any. So we decided to give it to the CSA folks with a *disclaimer*, and by and large, people were thrilled. It *was* the best corn of the season. . . . When we got our postcards back, one of the coolest things was one card that said, "More corn! More wormy corn next year! My kids have a worm habitat set." (See chapter 14 for more thoughts on quality.)

Offering a weekly bulk order, in addition to the prepaid share, will encourage your members to purchase more from you so that you have less need of outside markets. Harmony Valley calls this system "Shareplus." A bulk-order system will also help you avoid dumping more produce on your members than they really need. You can include in your bulk list produce rated as seconds, such as crooked carrots or nicked potatoes, for a lower price. For the GVOCSA bulk order each week at distribution I provide a list of items that will be available for the next week, with the price for each item. Members sign up for the amount they want and calculate what they will owe. The next week, those bulk items appear at distribution and members pay the coordinator. Since we do not supply winter shares, we try to stimulate bulk purchases for canning, freezing, and root cellaring by providing information on how best to store the food. Some members buy just a little extra through the season;

others stock up energetically for the whole winter.

The Food Bank Farm has invented another way to get members to buy more from them. They offer six different "value-added" packets with recipes. In Michael Docter's words, ". . . we sell a 'Pickle Package,' which consists of all the fresh ingredients needed to make pickles: cukes, garlic, onion, hot pepper, a bunch of dill, all prepackaged in a plastic bag with a recipe included. The package looks beautiful and people eat them up. We do the same thing with salsa, tomato sauce, pesto, and a number of other recipes" (*Growing for Market* [July 1996]: 6). You can purchase these "value-added recipe kits" from them.

If you do not have experience with other markets beyond your CSA, inform yourself thoroughly about market standards for quality and packaging. The nearest Extension office should be able to provide you with Federal and state standards for produce. With some markets, you can jeopardize all future sales by delivering your produce in the wrong kind of box. A restaurant will have different requirements from a retail store. Produce that might be acceptable to your CSA members or at a farmers' market stand may not look polished enough for the bright lights of a supermarket, where perceived quality is based on cosmetic perfection.

If you plan to sell your produce as "organic," you may need to have your farm certified. When implemented, the National Organic Program, adminstered by the USDA, will require certification for all farms with sales over $5,000 a year. Your state may have its own organic standards and enforcement mechanism. (See Chapter 11 for more information.)

CONNECTING TO THE EMERGENCY FOOD SUPPLY

Besides providing sliding scales for fees, accepting food stamps, and giving scholarships, many CSAs donate excess produce to the hungry. Where pickup is on the farm, delivering extra shares or finding a group willing to come for them on a regular basis may take some organizing. Likely candidates are local churches, meals-on-wheels programs, or caring societies that supply groceries or meals to the sick or infirm elderly. In some areas, county gleaning pro-

CLEMENS KALISCHER

Mia Rubow and a helper load boxes of cucumbers, lettuce, and tomatoes for wholesale delivery.

grams, or groups of gleaners organized by food banks or churches, will come to farms to pick or pick up food. The GVOCSA has used two churches as distribution centers. One gave the unclaimed shares to senior citizens who came to the church for meals; the other has a program for distributing packets to people in the parish. They decided to allocate the CSA leftovers to elderly people who appreciate the fresh vegetables and know how to cook them.

Winter Green Farm in Noti, Oregon, makes a special effort to contribute to the struggle against hunger in their community. First, their CSA, the Winter Green Community Farm, gives a discount of up to 20 percent for members who cannot afford the full price. Their contract with members solicits contributions to the Food for Lane County Program, which supplies area food pantries and other emergency food programs. In 1997 members donated over $1,500, which purchased six shares for participants in Womenspace, a transitional home for women escaping abusive family situations. One of the recipients expressed her gratitude: "I felt special that I could have fresh food. When I was down on my luck, it helped me through the hard times." The farm also gives unclaimed shares and excess produce in bulk to Food for Lane County, delivered by farm staff or CSA members (see chapter 20 for more information).

Gardeners know that some pretty ugly-looking food can be perfectly good to eat. When you get full from eating the factory rejects yourself, you need to find other hungry mouths to feed. While chefs at restaurants or institutional cooks may not want this food, people who run soup kitchens are often willing to take the time required to cut off bad spots. The county Extension office or local council of churches will usually be able to give you a list of soup kitchens and free meal providers in your area. The GVOCSA has a regular connection with St. Joseph's House of Hospitality, a Catholic Worker project that serves a daily noon meal to people in need. St. Joseph's is conveniently located a few blocks from the CSA pickup and is also willing to organize small crews to come out to the farm to glean.

◆

ON A FARM RUN WITH A HIGH LEVEL of ecological consciousness, nothing need go to waste. Top-quality crops will go for sale to primary markets, CSA or other. Processors or neighbors may provide outlets for culls. Crop that is not of saleable quality, but is still edible, can go to the emergency food supply system. Whatever is left can feed livestock, the compost pile, or the vast microherd in the soil.

REGIONAL NETWORKING FOR FARM PRODUCTS 18

As physical resources are everywhere limited, people satisfying their needs by means of a modest use of resources are obviously less likely to be at each other's throats than people depending upon a high rate of use. Equally, people who live in highly self-sufficient local communities are less likely to get involved in large-scale violence than people whose existence depends on world-wide systems of trade.

—E. F. SCHUMACHER,
SMALL IS BEAUTIFUL

THE SHARERS ASSEMBLED IN A CSA make a perfect group for bulk purchases of all kinds of sustainably produced items from other farms or enterprises. CSAs can serve as hubs for local or regional economic development. Berries, fruit, cheeses, wines, maple syrup, bread, eggs—even non-food products and services are possibilities. Linking up with CSAs can give a boost to the marketing of specialty farms, while members benefit from access to hard-to-find local products.

For these purchases beyond the regular share payment to go smoothly, however, some questions need to be answered. First, who is going to be responsible for choosing the products, and making arrangements for buying and delivering them? If the CSA farmer handles other people's products, how should the farmer be remunerated? Who sets the quality standards for outside purchases? What is the procedure if a product is not acceptable? Does payment go through the CSA treasury or directly to the outside producer? Who handles these payments and makes sure no one is shortchanged?

The GVOCSA has done some outside purchasing for a few years and tried out a variety of approaches. Our experience highlights some of the problems that can arise. In 1993, I suggested that members might like to try getting chicken from Backbone Hill Farm. A member of the core group agreed to be chicken coordinator. Beth Rose supplied us with her price list and the coordinator took orders. (A few vegetarian members were mildly put off by a meat deal.) Beth then delivered the chickens to Rose Valley, where we kept them in the walk-in overnight for members to deliver to Rochester the next day together with the week's shares. Rose Valley's part in the deal was small, but Beth gave us a few chickens for our help in connecting her with a new market.

The first year, Beth brought the chickens in large boxes on which she wrote the bulk weight. The chicken coordinator then weighed out the chickens one by one for the orders. As one might have predicted, the two scales did not agree, or the smaller quantities did not round out exactly to the bulk weight. Whatever the cause, there was some trouble over exactly how much the members owed Beth for the chicken, and I was the apologetic intermediary. The next year, Beth switched to weighing each chicken herself and making deliveries directly to Rochester. The GVOCSA chicken-eaters and I regret Beth's passing. Her surviving partner, Kim McKnight, keeps the chicken business going, though he no longer markets to Rochester.

With another neighboring farm, we arranged for bulk purchases of organic strawberries with the farmer delivering the berries to our walk-in for transshipment. His first berries were beautiful, top quality. In the second batch, however, I noticed bruised berries. Yet I did not want to tell my neighbor to take them home. No one had put me in charge of his quality control. The distribution coordinators ended up selling some of these berries for less than the agreed price, and the farmer returned the money to members who paid the full price but complained. When we discussed the incident at the core meeting, we decided that, in the future, the quality decision would be up to the distribution coordinator.

Since the strawberry disappointment, the GVOCSA core has handled all outside purchases directly. A member volunteers to be coordinator for a given product and takes charge of sign-ups, communication with the producer, and the collection of payments. Pat Mannix, an active member of the core since year two, has taken charge of organic wine and maple syrup. Her observations are instructive:

> When they pick up their shares, everybody gets a flyer that tells what's available and the price. Then we have a sign-up sheet at distribution, and there'll be a clipboard and an envelope for payments. The sheet tells who to make the check out to, and that you have to prepay, and when the stuff will come, but nobody ever knows the answers to any of those questions. I go and I pick up the sign-up sheet, and there will be the names of people who have no checks in there, and checks from people who haven't put their names on the sign-up sheet, and people who add up what they bought and it'll be $12 short or $4 over, and cash in the envelope, when it says "No Cash Allowed." It's a wonderful thing. Then you get to know the members because it says name and phone number, and they never put their phone number down, and when you look at the membership list to see what their phone number is, then they're not on the membership list. So you think, where did they come from and why are they on the list? Really, it's a fun thing to do!

Though she will tell you that communication is not really possible, Pat has stuck with this project for the full ten years. She says that the alternative to leaving the sign-up sheet is to sit at distribution *with* the sign-up sheet, which results in greater accuracy, but takes much more time. Due to Pat's efforts and a few others, GVOCSA has purchased organic wine, maple syrup, sheep cheese, and low-spray and organic apples. Our members get the price advantage of bulk orders, and we farmers get to mind our own business!

In 1997, GVOCSA invited Jim Pecora, who raises free-range chickens to offer chicken and egg shares. He delivered weekly eggs, and monthly frozen chicken directly to the distribution center, leaving them packed in ice in Styrofoam boxes. This system worked smoothly for everyone. Besides including Jim's flyer in a regular mailing, the core had little involvement.

Two of the branches of the Roxbury Farm CSA are trying systems of their own. According to Marcie Shemaria, a member of the Columbia County core group that organizes distribution at the farm, in 1994, the core decided it might be nice to sell surplus cheese from the nearby Camphill Village. They set up an order system, including dairy products from Hawthorne Valley Farm, also nearby. Members order every week for products delivered the next week. This project has blossomed into Community Agriculture of Columbia County, a pre-order co-op with its own bank account, tax number, and membership fee. Around 120 of the 165 households in the CSA have joined. The co-op charges a 20 percent markup to cover costs and deals with twenty-one different suppliers producing a wide array of products, including eggs, pesto, cheeses, honey, maple syrup, gazpacho, and bread. In place of the "Shareplus" system Roxbury tried out for a few years, it too has switched to selling bulk vegetables for canning or freezing, plus sauerkraut and meat through the pre-order co-op. The co-op's efficiency got a big boost when they adopted a computer system called "Provision," which is available free of charge from Northeast coops. Provision keeps track of each member, produces an order form for each of the suppliers, a distribution sheet so that core members can put the correct items in each member's box, and an invoice for each member. It can also make labels. Processing forty orders a week takes three to four hours of computer time. Members volunteer time to help run the co-op, but the people who coordinate and keep the books on a regular basis are paid with a $5 an hour credit to the co-op.

Brookfield Farm
Amherst, Massachusetts

The extra purchase system of Roxbury Farm's New York City branch is not as ambitious. In 1996, to help make the members feel more connected to the CSA and support a local business, the core started to offer bread from Amy's, a New York City bakery. As core member Jennifer Castle tells it, they kicked sales off with a little bread-tasting party at distribution. The bread made a big hit, and members began ordering once every three weeks for the next two weeks. Jennifer took the orders, counted the money, called the orders in to Amy's, and made sure the checks were deposited and that the treasurer sent payment to the bakery. All this took too much of Jennifer's time, so the next year she split the job with another core member. Together they added maple syrup and honey, allowing members three weeks to order, but only one to pick up. Selling fruit shares for the winter, they discovered how much easier it is to handle one advance order for the entire season. In the future, they plan to sell bread shares this way too, in three-month segments. Any food not claimed on time goes to a food cupboard that operates out of another part of the church where the pickup takes place. Shares of vegetables and bread compensate the core members who do the coordination work.

Instead of the sharers making bulk purchases as a group, in some CSAs the farm does the buying and sells to the members. Michael Docter's Food Bank Farm came up with a scrip system to simplify the bookkeeping. For $10, the farm sells a "Scrip Card" with five $2 punches on it. They divide the produce into $2 packets: 3 pounds of pears, 4 pounds of apples, or 5 pounds of potatoes for $2 each. According to Michael, they purchase field-run produce, unsorted and ungraded, at a discount, and resell it to members for a good price while maintaining a profit margin of 20 to 40 percent. Each scrip card is numbered, and a perforated section is torn off at each sale to provide the farm with a receipt.

Dan Kaplan of Brookfield Farm, near Amherst, Massachusetts, buys products from other farms and resells them to his CSA members, who all pick up at the farm. He adds a small mark-up to cover his crew's labor and they literally eat or drink the leftovers. Financially it is a wash, but Dan considers the additional products a useful service both to members and to the other farms.

WHETHER A CSA sells other farms' products as supplements to its own production, additional shares, on-farm purchases, or through bulk or cooperative sales depends on the personalities and individual needs of all involved. In developing these markets, it is important for CSAs to be aware of other struggling local sustainable businesses, such as food co-ops, so as not to become competition where cooperation might be possible. The Genesee Food Store, a co-op in Rochester, New York, finds that, whatever it loses in vegetable sales to the GVOCSA during the summer, it more than makes up in additional highly food-savvy members in the long run. The success of efforts such as Community Agriculture of Columbia County offer hope to all of us who want local food systems to replace the global supermarket.

— Part V —

Many Models

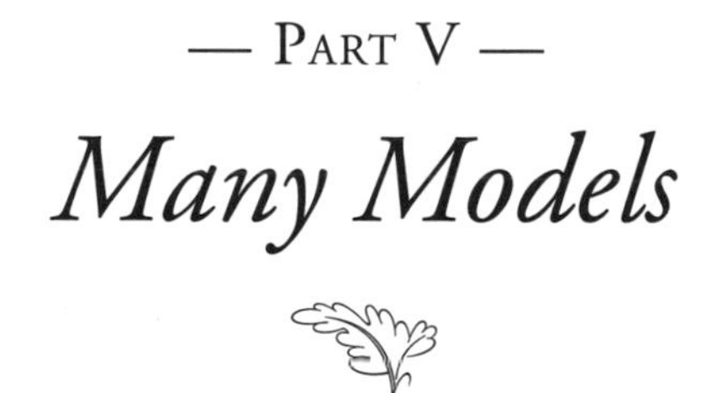

Multifarm CSAs 19

Farmers are pitted against one another in the prevailing system of commodity production, in ways that work to the detriment of all . . . a new production ethic will emphasize solidarity, mutual support, and interdependence among farmers.

—Vision Statement/Call to Action for "The Soul of Agriculture: A New Production Ethic for the 21st Century," 1997 conference

Juggling thirty, forty, even fifty crops is not to the taste or within the technical capabilities of every farmer. Some groups of farmers have chosen to associate to produce the crops for a CSA. Consumer-initiated CSAs have also combined crops from a few farms instead of focusing on only one. There are advantages to sharing the work and sharing the risk, but coordinating harvesting schedules and deliveries, apportioning crops fairly, and agreeing on quality standards are organizing challenges each group must overcome. Once a solid group of farms has formed, however, the possibilities for additional cooperation—joint purchases, sharing equipment, developing other markets—are limitless. Networks of small and medium-sized farms, whether farmer, consumer, or organizer instigated, can become the backbone of a sustainable local economy.

In her keynote speech at the Pennsylvania Association for Sustainable Agriculture (PASA) conference in 1996, Sarah Vogel, then commissioner of agriculture in North Dakota, made a very significant distinction between "economic development" and "development economics." Economic development is the familiar phenomenon: towns, counties, states, and even nations compete in offering incentives to entice enterprises from outside to select their community. Development economics works differently: it begins with an assessment of the existing resources of a community, and creates an economic strategy to strengthen and build on those resources. In North Dakota, the Department of Agriculture funded a feasibility study for a pasta cooperative proposed by the state's wheat growers. The cooperative pasta factory has become a thriving enterprise, adding value to the raw commodities produced on grain farms of the region. Forming multifarm CSAs is development economics applied to smaller, more diversified farms.

FARMER-INITIATED GROUP CSAS

Lynn Byczynski and Dan Nagengast, renowned in the small farm world as editors of *Growing for Market*, started a CSA on their own farm near Topeka, Kansas, then decided to expand it by inviting other farms to join them. In the winter of 1993–94, with seed money from a Kellogg grant, the Rolling Prairie Farmer's Alliance formed as a farmer-owned cooperative in Lawrence, Kansas, which has grown in membership every year since. In 1997, eight farms supplied the produce for 325 shares; for 1998, the co-op had a goal of 350 shares. Half of the farmers have had extensive experience selling at farmers' markets. None of them is totally new to marketing.

Rolling Prairie Farmer's Alliance
Lawrence, Kansas

In December, the farmers get together to decide on the crop mix for the coming year, based on surveys of the members. Each farmer makes a commitment to the dollar value he or she wants to sell through Rolling Prairie. Most of the farms have other markets as well, so Rolling Prairie is not under pressure to absorb all of members' production, and there is plenty of backup supply. Who grows what is an evolving process, according to member Paul Johnson, of East Stone House Creek Farm. He began by selling strawberries through Dan and Lynn's CSA, and has expanded to many new crops, including asparagus. The farms share responsibility for the harder-to-grow and labor-demanding crops. The co-op pays one farmer to act as produce manager, which entails calling all the other farms each week, deciding what goes in the weekly share, which farms will provide it, and calling in the orders to the farms. The manager aims for a weekly bag value of as close to $11 as possible. Several farms often split crops, such as peas and green beans. The manager is also empowered to do quality control, but so far has had to pull out only a few items. The co-op asks its members for their best quality. Occasionally, because of deficiencies in supply, the co-op cuts the value of the bag for the week. To fill out the shares, the co-op purchases asparagus and apples from other farmers.

In 1997, Rolling Prairie charged $11 per week for twenty-five weekly bags of fresh produce. The co-op's accountant, also a farmer member, uses local retail prices as a guide to determine payments to farmers. To help cover some of its administrative costs, the co-op also buys a few crops wholesale, and resells them to consumer members at retail price. Farmer members share most co-op tasks, but the group pays the produce manager, the accountant, and the newsletter writer. Consumer members pay an annual deposit of $35: $10 for administrative costs and the newsletter, and the balance applied to their final month's produce bill. The co-op bills them monthly; members pay after receipt of the food. Besides the weekly share, the farmers also sell organic chicken, eggs, beef, lamb, frozen raspberries, bulk quantities of produce for canning or freezing, and cut flowers.

Twice a week, farmer and consumer members converge on Community Mercantile, a natural foods cooperative store in Lawrence. The farmers, who arrive an hour in advance, set out boxes of their produce on a counter in the deli section with signs telling members how much to take of each item. Members have from 4:00 P.M. to 7:30 P.M. to pick up. Anything left over goes to emergency food providers—Shalom House in Kansas City or Achievement Place in Lawrence. Synergy between the two co-ops seems to benefit everyone; Rolling Prairie has brought new members to "the Merc" and many sharers do other shopping when they come for their bags. Merc nutritionist, Nancy O'Connor supplies weekly recipes for the newsletter, and the deli cooks samples for members to taste. Paul Johnson describes the deli manager as "a gem," eager to feature the dishes from Nancy's recipes and to use Rolling Prairie food.

Rolling Prairie's goal is to make buying from local farms as easy and attractive as possible. The co-op unabashedly follows the subscription model for its CSA, asking consumer members for no participation beyond purchasing shares. The cooperative efforts of the eight farmers greatly reduce the risk to sharers of crop failures or shortages. Sharers only pay for what they get. For the consumers, the farmers try to build a sense of community and provide education through their newsletter, as well as farm, garden, and greenhouse tours, flower walks, and potluck dinners. For the farmers, the close working relationships with each other have value beyond the business connection, enabling them to share friendship and information, equipment, such as a tub and grinder for chicken feed, cooler space, and even fields. (Land-poor farmers are growing crops on the larger farms.) In the future, they plan other joint purchases, and Dan Nagengast has received a SARE grant to start a processed tomato co-op, which will include many of the Rolling Prairie farms.

During its first few years, the Organic Kentucky Producers Association (OKPA) has had more trouble building trust and cooperation than Rolling Prairie, but seems to have arrived at a committed group of farmers who can work together. For 1998, eight farms, all certified organic by the State of Kentucky, will be supplying one hundred shares to families in the Lexington area. OKPA got its start in 1993 as an offshoot of the Commodity Growers Coop, a farmer marketing organization established to find alternatives to tobacco to enable small Kentucky farms to survive. The Jenkins family, featured in the Public TV program "Tobacco Blues," is a member. Organized as a farmer co-op, OKPA has four member farms who own the co-op, and four associate members who sell to the co-op but lack voting rights. After two years of contention over favoritism and a suit against a member, OKPA's main goal for 1998 is to get stabilized. Unlike Rolling Prairie, OKPA collects payment for its shares in advance, with a choice between Plan A for $492 and Plan B for $300. Pricing is based on retail prices, aiming for the high end because of quality, freshness, and service. Most of the members prefer home delivery, for which they pay $2.50 a week. The co-op pays one of its farmers to do the deliveries. According to co-op farmers Pam and Lawrence Jenkins, most members are upper-class doctors, lawyers, and professionals.

Much of OKPA's process resembles that of Rolling Prairie. The farmers meet to decide on crops. A coordinator designs the weekly shares and communicates with the member farms. Each farm is responsible for its own quality, but all the farmers gather at one farm to pack the boxes for delivery and to look over everything and come to a consensus about problem crops. The coordinator has the final say. OKPA would like to expand beyond vegetables to other products, such as eggs, baked goods, and jams and jellies.

Very few groups of farms seem to be attempting to form integrated communities like Temple-Wilton Community Farm, where three farm entities merged into one. Several farms with established CSAs, however, serve as the market for a cluster of other farms in their area. Roxbury Farm in New York hosts Community Agriculture of Columbia County, a preorder co-op carrying products from over twenty regional farms; Silver Creek in Ohio sells shares of neighbors' beef, and goat's-milk cheese; Peach Valley Farm in Colorado assembles shares of a wide assortment of fruit from half a dozen neighboring farms. Three farms in New Hampshire have associated as North Country Community Farms to offer their forty-three members "Early Greens" shares from one farm, apple/cider shares from another, and the choice of "Organic Garden" shares and "Putting-Up" shares from two of the farms. As family expansion, financial pressures, and other factors have fluctuated, the core group has taken an increasingly active role in making flexible arrangements with the farmers and adjusting members' work schedules to farm needs.

AGRICULTURAL DEVELOPMENT

In several parts of the country, organizations formed to reconstruct the rural economy have chosen CSA as one of their tools. Closely related to (or even overlapping with) community food security efforts, these projects combine support for local farms with the creation of job opportunities and access to improved nutrition for low-income residents. In the Appalachian area of southwest Virginia and northeast Tennessee, the Clinch Powell Sustainable Development Initiative (CPSDI) shepherds a series of projects to create more local businesses in sustainable wood products, sustainable agriculture, and nature tourism. The CPSDI brochure proclaims, "The more people that are involved in local business activity, the more local people make money, the stronger the local economy becomes, and the more we get for our own natural resources. It's a simple idea. It's called 'value added' economic activity. *Value added.* It makes good sense for people *and* the environment." So far, organizer Anthony Flaccavento has facilitated the birth of two CSAs and has plans for more.

The Highlands BioProduce Cooperative, one of CPSDI's projects, coordinates the sales from a dozen small farms to a CSA of one hundred shares, a network of restaurants, and value-added products packaged in "Appalachian Harvest" gift baskets and "Pecks of Pretty Peppers." Half of the farmers work full-time at farming; two raise beef; three are Amish family farms. One of the farms functions as the "block house," where the farmers assemble weekly share bags and loads of produce for restaurants. The share price of $360 for twenty-six weeks saves sharers 5 to 7 percent over a comparable packet from the local Kroger's supermarket. The one hundred sharers are divided into clubs for distribution; three groups use churches and one picks up at the central farm. Only a few sharers purchase working shares at $320 with a ten-hour work requirement. The project has focused most of its energy so far on building farmer participation, marketing, and biological farming skills.

The second CPSDI-initiated CSA is not multi-farm, but high-school–based. In Sneedville, Tennessee, a town of seven hundred in Hancock County where 40 percent of the people live below the poverty line, CPSDI's Jubilee Projects decided to try an organic CSA on school land. They hired Kyle Green, a local farmer who is skilled at vegetable production, but had never used organic methods, to teach the high school students to run the CSA. The project has two years of funding before it must become self-sufficient. During the first year, 1997, thirteen families participated.

In Story County, Iowa, where 990 farms cover 90 percent of the land and gross sales of agricultural products topped $15 million in 1994, the organizers of Magic Beanstalk CSA have had a hard time finding enough skilled vegetable growers. As farms have gotten bigger, Iowa's rural communities have declined. The poverty level, though not as severe as in Appalachia, is a distressing 9.2 percent. The Magic Beanstalk is one of a number of projects stimulated by the Kellogg-funded Shared Visions program to rebuild the local economy. The story of the Magic Beanstalk goes back to 1993 when Shelly Gradwell got snowed in at the NOFA-NY Conference, where Katy Sweeney gave a workshop on Malven Farm CSA, and we spent the extra day in an impromptu discussion on breaking new ground. Shelly brought the CSA concept to her graduate studies and Extension work in Iowa. In 1995, Mark Harris grew one acre of vegetables, supplying the thirty shares for Magic Beanstalk's first season. That year, Iowa saw its first three CSAs. By 1998, twenty-seven were underway and ten more were in the planning stage.

For 1998, Magic Beanstalk's strong core of organizers has recruited three farms to supply 150 shares. Iowa Fresh Produce, Onion Creek Farm, and Heenah Myah Student Farm have divided the crops, with the more experienced farmers at Iowa Fresh growing the largest number. (Gary Huber, who also works as staff person for Practical Farmers of Iowa and its

Magic Beanstalk
Ames, Iowa

Field to Family Project, farms at Onion Creek). The student farm, a new undertaking, grows the labor-intensive peas and beans, as well as pumpkins, on land belonging to Iowa State University. The share price is $250, with members urged to pay in advance. Field to Family, the offshoot of Magic Beanstalk funded by Community Food Projects, provides subsidies for families who cannot pay that price (see p. 208 for more details). Magic Beanstalk's tastefully designed brochure emphasizes community and member involvement. The front panel makes its mission clear: "Providing central Iowans with fresh, locally grown, wholesome farm products. Bringing people together to recognize and celebrate the important role good food and local farms play in a thriving community." The contract section underlines the importance of member participation, and solicits help with farm work, food distribution, administration, newsletters, education, and outreach: a core group of twelve organizes all of this activity.

In addition to vegetable shares, Magic Beanstalk also offers a wide assortment of products available at the CSA distribution site, including beef, pork, lamb, chicken, turkeys, eggs, garlic, honey, fresh bread, apples, cider, whole grains, raw yarn, woven goods, cut flowers, and okra and black-eyed peas from twelve other farms. The last come from "Soul Food on the Prairie," an organic farm. Each farm has a brief write-up; some profiles describe the farms as organic, others as growing without chemicals or antibiotic-free. The twelve include Dick and Sharon Thompson, national leaders in on-farm research, and Denise O'Brien, past president of the National Family Farm Coalition and eloquent spokesperson for the cause of family farming.

Suffering from an excess of tourists but poor produce sales, the Pike Place Market in Seattle, Washington, decided to stir up business with a multifarm CSA. The market hired Linda Bartlett to manage the CSA; provided a cooler, staging area, and delivery vehicle; and subsidized the administrative expenses. Linda recruited members in an office building near the market and four reliable farmers as core suppliers. To help improve the market, the farmers had to agree to participate one day a week. The group combined farmers from eastern and western Washington, with their dramatically different growing conditions, for maximum variety and seasonal range of crops. Linda purchased some items from a dozen other farms to make an attractive share packet. Only twice over the course of the twenty-two-week season did she exercise her right to reject produce. The price to consumers was $23 a week, of which $18.50 to $19 went to the farmers, $2.30 to Linda, and $1.50 for delivery. The CSA paid the farmers retail prices minus 10 percent discount, with 25 percent up front, and the rest cash on delivery. When Linda calculated her hourly pay, a meager $3, she concluded that the administrative fee would have to rise to 20 percent to make the project sustainable. Although basically a subscription service, the CSA worked at establishing relationships between members and farmers through a weekly newsletter sent via office e-mail, and through festivals held at two of the farms.

CONSUMER-INITIATED CSAS

"What do you do if you live in a suburb where there *are* no farms?" asked Sarah Lincoln-Harrison and her husband Richard "Pic" Harrison. Their answer: they organized the Marblehead Eco-Farm in Massachusetts as a charitable corporation and started a

CSA. (See p. 27 for the story of its birth.) Farmers hungry for markets in other parts of the country will shake their heads at the irony of how hard the Harrisons had to hunt to find a suitable farmer. In 1998, its sixth year, the Eco-Farm CSA will be buying crops from four farms to provide shares for 130 members. A steering committee of seven, including the Harrisons and a farmer representative, oversees the project. Eco-Farm pays farmers the best price they think the members will accept, roughly 80 percent of the retail price in an upscale produce market. Members make five payments from December through August, but farmers receive monthly paychecks year round, with amounts based on the prior year's sales. The price of a share is $260, plus a $40 membership fee to cover administrative costs, a 170-page cooking resource guide, and a weekly newsletter. Members can also purchase a fruit share for $75, and extended-season winter crops for $45. Eco-Farm offers $100 scholarships for working shares, and donates leftover food to the Crombie Street Mission in Salem, over $6,000 worth in 1997.

Eco-Farm pays farmers the best price they think the members will accept, roughly 80 percent of the retail price in an upscale produce market.

Sarah describes the members as "representing a broad spectrum of understanding and commitment, from those who simply want fresh, local produce to those who desire above all to obtain organic food and to support the viability of regional farms." Therefore, the group's policy is "not that purist." Their primary farmer uses organic practices, but is not NOFA-certified. A second organic grower delivers potatoes, carrots, and winter squash on a monthly basis. In 1998, a fourth grower was brought in to cover deliveries during the heavy spring rain. The new Eco-Farm manager coordinates the weekly share menu for members from the multiple farms.

Until 1998, the Harrisons and treasurer Don Morgan did all of the administrative work for Eco-Farm, Pic putting in fifteen hours a week on office tasks and Sarah working fulltime to the CSA and the group's other projects. The bookkeeping system Pic uses is Microsoft Access, and he is willing to share his pricing and payment formulas. In 1998, they hired Jennifer Mix as full-time manager and purchased a van for her to use to relieve the farmers of the chore of delivering to Marblehead. She will act as manager for both the CSA and the Saturday farmers' market, which Eco-Farm initiated with a one-day festival in the fall of 1997, drawing a crowd of two thousand. The plan is for one-third of Jennifer's salary to come from farmers' market fees of $25 per stall per week and from the sale of organic shade-grown coffee and value-added products at the Eco-Farm booth. Two working sharers will continue to help overseeing distribution at "the depot," a shed constructed by the project on town land.

Sarah says the CSA has become less intimate as it has grown to 174 families in 1997, and members are having a harder time grasping the notion of sharing the risk with the farmers. With the advent of the farmers' market and a new option for those seeking organic produce on a more selective basis, enrollment for 1998 dropped to 130 families. However, comments from 1997 participants suggest a high level of enthusiasm.

"I'm glad we signed up."

"I've discovered I love the kohlrabi."

"More, more, more!"

"Too many greens! But we're learning!"

"The late start was unfortunate, but it was worth the wait! Really appreciate the farmer's treats, and the fruit has been outstanding. The organization of the depot seems to be running really well."

"Every year it gets better. Easy to identify fruits, vegetables, and farmer's treats. Layout facilitates quick pickup and also room to socialize. The weigh stations are well positioned. Beautiful planted trees provide welcome shade and remind us of our connection to nature."

Although not all of the members pitch in, many spend a few hours helping out at the depot or staff the Eco-Farm booth at the farmers' market. Taking part in the CSA has led members to become more active in other Eco-Farm projects—an organic gardening course, as well as other school programs and farm tours. Eight member families will be growing an heirloom variety of pole beans for seed. "Once you participate on the fringe of sustainable agriculture," Sarah explains, "it opens you up for deeper involvement." Sarah hopes that her work will contribute to making life on this planet more sustainable. At the very least, the Eco-Farm CSA, Sarah says, "has brought balance to a community that is heavily involved in soccer and sailing, and fosters an appreciation of the land that continues to disappear with indiscriminate development."

Marblehead Eco-Farm
Marblehead, Massachusetts

Farther to the north and west, a group of eighty city families in Ottawa, Ontario, runs a CSA that they call "OOFA," the Ottawa Organic Food Alternative. OOFA member Randi Cherry described this project at the Guelph Organic Conference in January 1998 as a way for non-farmers to "share in the discipline and fun of this 'seed to supperplate' aspect of their lives." The Ottawans began with ten families buying vegetables from one farmer in 1988, and have expanded to purchases from over twenty farms that supply vegetables, fruit, breads, goat's milk, soya products, honey, maple syrup, hempseed cookies, wild rice, and blueberry spreads—all of which "feed our bellies and nourish our souls." The members gather on Sundays to divide up and pay for their orders. By ordering on a weekly basis, they are able to buy exactly what they want, but the members do all the work of collating orders, collecting payments, and picking up from the farms. OOFA is an active group: members pay $52 a year, do thirty-six hours of volunteer labor, work a day on a farm, and add 15 percent to the produce costs to cover group expenses. They are also willing to lend seed money or raise funds for the farms from which they buy. Randi's concluding words bear repeating:

> The core issue for our CSA is nourishing the growth of self-sovereignty so that our members increasingly choose to transform fear and isolation into love and community. In true community, wilderness can expand throughout cities, farms, and natural spaces. Many groups of people—metropolis large or village small—are currently choosing daily lifestyles and environments that sustain all life forms. CSAs are one important means to this end. ("Guelph Conference Proceedings," pp. 34–35)

LIKE THE OLD ARGUMENT among actors about whether you achieve the most genuine emotion on stage by reliving the feelings from the inside or by imitating the outward gestures (a great performer can get there either way), CSA can arise from the collective efforts of farmers reaching out to consumers, or consumers reaching out to farmers. Some projects use financial incentives to get people to work; others rely on volunteer energy. The revival of local food systems through agricultural development economics is the goal. As with genuine emotion, we will recognize it when we get there.

MATCHING BIODIVERSITY WITH SOCIAL DIVERSITY

20

Everyone does better when everyone does better.

— JIM HIGHTOWER,
THERE'S NOTHING IN THE MIDDLE OF THE ROAD BUT YELLOW STRIPES AND DEAD ARMADILLOS

A POWERFUL STREAM OF CONCERN for social justice runs through the movement for a sustainable food system. In "Peace with the Land, Justice among Ourselves," Marty Strange suggests that "the growing income inequality that divides our world" is the greatest threat to ever achieving a truly sustainable agriculture. He asks: "Would we be comfortable with a dual food system in which the rich paid a premium for food produced by agronomically wholesome means, while the poor ate cheap food produced by making war on the land?" Recoiling in distaste from the notion that organically grown food is exclusively for "yuppies," many CSA projects are seeking creative ways to balance financial support for their farmers with including members who have little money to spend on food. Community organizers eager to empower the poor to take charge of their own food supply see the potential of CSAs to involve and train people as active members and staff.

SUPPLYING EMERGENCY FOOD

The simplest way to make sure that at least some of the CSA food reaches low-income people has been to donate leftover shares to food pantries or soup kitchens. CSAs that do their distribution in churches often find a ready connection with existing food programs. Parishioners in the Lutheran Church of Peace deliver leftover shares from the GVOCSA to elderly members of the congregation. When boxes of food remain at the end of the agreed-upon hours, site hosts for Angelic Organics call other members who have volunteered to take the boxes to food pantries. If sharers have to miss a week, many CSAs give the food to the hungry.

Taking a much more ambitious approach to getting fresh produce into the emergency food supply, some food banks have established their own farms.

The Food Bank Farm of Western Massachusetts (I tell its birth story on p. 27), under the energetic leadership of Michael Docter and Linda Hildebrand, serves a CSA of eight hundred families with five hundred shares, while supplying an equal amount of food to its parent food bank. In 1997, the farm produced 180,000 pounds of food for agencies supplied by the Food Bank. Only a tiny percentage of that passes through the Food Bank warehouse: the agencies pick up directly from the farm three days a week. Since the farm has no refrigeration, Michael and Linda don't grow many tender greens for the agencies, concentrating instead on "hardware" such as carrots, cabbages, and summer and winter squash. They also grow produce for the Food Bank's brown bag program, supplying the right number of units directly to the various distribution sites. The farm gets some volunteer labor from Food Bank supporters, but does not make training recipients of the food in farming skills a priority. Until 1997, Michael and Linda had to balance dual roles as farmers and employees of the Food Bank. Tensions resulting from missed staff meetings when farm work took priority led them to negotiate a new structure with the Food Bank: Michael and Linda now have a contract to run the farm autonomously and pay for their lease with 40 to 50 percent of production plus some cash.

In selling its shares, the Food Bank Farm stresses its social mission. Their brochure carries this message:

Your share of the harvest provides nutritious foods for:

- the unemployed family at a Springfield soup kitchen
- the battered wife in a shelter in Greenfield
- the mother feeding her family from the Northampton Survival Center
- the hilltown widow trying to get by on a social security check.

Michael insists, though, that it is the economically produced high-quality food that retains members once they join.

Other food bank farms have not yet managed to be as impressively productive, and a few have been total flops. A food bank in upstate New York tried to benefit from a piece of muckland that a well-wisher allowed them to use free of charge. Unfortunately, the free land did not come equipped with a free farmer, farm implements, or any infrastructure. Not too many of the potatoes that grew made it out of the ground and into hungry mouths.

The Greater Pittsburgh Community Food Bank sponsors the Green Harvest program, which has an ambitious list of interrelated goals: to build local support for local farms and urban gardens, to eliminate hunger, and to encourage self-reliance. Green Harvest sponsors several farm stands in inner-city neighborhoods, recruits experienced gardeners to train new participants in the urban gardens it has established, and coordinates volunteers to do gleaning and farm work on the farm it has been trying to consolidate since 1992. The farm began with a lease arrangement with an independent farmer, who then turned over his land with its CSA to the Food Bank. With only four years' experience in farming, Mike Gabel took over as farm manager of the Green Harvest CSA in 1997, the year the farm had to move to a new piece of land. Largely due to the new location, CSA membership dropped from 110 to 25. Taking the Food Bank Farm of Western Massachusetts as a model, Mike's goal is to reach self-funding through the sale of shares and to supply half the farm's production to the Food Bank. After one season at the new site, shares covered 50 percent of the farm's funding needs. The many other social goals of the Green Harvest project complicate the already challenging goal of increasing food production on the new farm.

The farm gives the Food Bank increased control over the fresh produce supply, good publicity, lots of work for volunteers, and a forum for relating sustainable agriculture and hunger issues. At a Pennsylvania Society for Sustainable Agriculture workshop, Mike delivered this impassioned statement on the need for clarity in the relationship between the food bank and the farm, which would apply with equal force to any other institutional farm arrangement:

> You need to tell the food bank realistically what kind of support you are going to need. . . . If you are to be self-sufficient, you are going to have to reach a certain threshold of CSA members to cover the cost of that farm, plus a surplus, a *big* surplus. It's critical that you and the administration of this organization are on the *same* page. You also need to decide, along with the food bank, whether your operation is going to have an educational component or whether it's going to be purely production in pounds. If you have all these people coming onto your farm and you are doing educational tours, who is planting stuff when it needs to be planted? Who is dealing with those groups? A lot of funders want you to do Herculean things. They want you to produce big pounds, they want you to have the most incredible, all-inclusive educational component that ever existed, reaching every inner-city kid and, by the way, why don't you put numbers on it? You've got to get your expectations straight with the administration and the farm—what is this beast going to do? Don't even put the plow in the ground until these things have been ironed out!

Kristen Markley's master's essay on "Sustainable Agriculture and Hunger" (Penn State, 1997) documents the experience of five Northeast nonprofit organizations, four of them with CSAs. Committed to both alleviating hunger and supporting small farms, Kristen provides a realistic appraisal of the pressures and obstacles facing small farmers and hunger organizations, and the benefits and difficulties of working together. She interviewed farmers employed by the nonprofits and independent farmers who sell to them, as well as office and management staff. She observed a high level of turnover among farmer-employees. She supports Mike Gabel's plea for clarifying priorities and setting realistic expectations. If a food bank expects young, inexperienced farmers to produce large enough amounts of diverse crops to make a farm financially self-sufficient, while at the same time training and empowering low-income community members to produce food themselves, the frustration levels on all sides are likely to be high. Greater familiarity with farming among nonprofit staff members, and better training and higher salaries for farm staff will help resolve these difficulties.

INCLUDING LOW-INCOME MEMBERS

In addition to giving away food, CSAs have sought ways to include people of diverse income levels and ethnic backgrounds among their members. One reason GVOCSA decided to ask all members to work was to keep the share price low. We also offer a sliding scale for payments, emphasizing that those who pay on the high end are subsidizing those who pay on the low end. Rose Valley Farm accepted food stamps, and the GVOCSA treasurers are willing to adjust payment schedules to accommodate people with low incomes or to satisfy food stamp rules. In its third year, the GVOCSA set up a scholarship fund to support membership for people who could not even pay the lowest rate of the sliding scale. The Downtown United Presbyterian Church generously provided grant money for the fund, which was also fed by members' donations and sales of the *FoodBook*. Despite all these measures, and strong efforts by Alison Clarke, the staff person for the Politics of Food, only a few inner-city, low-income people have stayed on as members for more than a year or two. Each person has a different story—either the food did not suit family tastes, or a combination of family, health, and job disasters made a steady weekly commitment impossible. Middle-class people with a job or a career and health insurance find it hard to imagine the upheavals the poor suffer in their daily lives. The most effective way we have found to support diverse members is to buddy them up with other members who have reliable automobiles and more stable lives. Genuine friendships are the strongest basis for retaining members.

Regional support organizations such as the Madison Area Community Supported Agriculture Coalition (MACSAC) and NOFA-VT are supplementing the efforts of individual CSAs. With funding from the Kellogg Foundation, the Partner Shares Program in Wisconsin is subsidizing shares for low-income community groups. As with the GVOCSA, everyone pays something, if only a few dollars per week. Sharon Lezburg, the program coordinator, helps find the groups and link them with existing CSAs. As the program reaches out to new ethnic groups and organizations it has achieved a 60 percent retention rate in the first two years. John Greenler, the MACSAC general coordinator, reports the program is working well with an African-American group and a battered women's shelter, but Hmong and Ukrainian community groups have not taken hold. Money for Partner Shares comes as well from the "Empty Bowl Project." At benefit dinners, potters and members of Chefs Collaborative 2000 combine crafts to sell bowls full of gourmet food with the proceeds going to Partner Shares. Sharon Lezburg is also doing outreach to church congregations that have organized "farmathons," similar to crop walks or marathons except that, instead of walking or running for money, volunteers do farm work.

Since 1994, NOFA-VT has been coordinating Vermont Farm Share, a program to enable low-income people to eat more fresh produce, learn about growing and preparing food, and connect with local farms by joining CSAs. In partnership with the Vermont Anti-Hunger Corps, local social services, and the University of Vermont's Cooperative Extension, Farm Share identifies appropriate families or centers, links them with a local CSA farm, and provides partial payment for the shares. By late 1997, 150 families, four community centers, and thirteen farms had participated. To raise money, NOFA-VT solicits donations from individuals and holds an annual Share Our Harvest event, in which people dine out and the restaurants contribute a percentage of their proceeds from the meals. During the first two years, before Kate Larson took over as staff person, Farm Share paid the entire fee for a full share for each family. As a result, Kate reports, a lot of food was wasted. In 1997, the program asked the families to pay on a sliding scale, and grouped them so they could divide both large-size shares and the responsibility for pickups among two or three families. Starting with smaller quantities of food made membership easier. Even with a subsidized fee, lump-sum payments were beyond the means of most of the families, so the program asked the farmers to arrange more flexible payment schedules.

Green Harvest CSA
McKeesport, Pennsylvania

Food and farm education is one of Farm Share's goals, but since low-income people must so often jump through hoops to receive services, Kate does not want to require formal classes. Farm Share does offer workshops on topics requested by participants. Kate hopes that learning will also happen naturally through the CSA process with recipe exchanges, newsletters, and farm visits. She stresses the importance of finding the right balance between giving support to families unaccustomed to eating a lot of fresh vegetables, while treating them as regular CSA members so that they feel a direct relationship with the farm. Kate did extensive telephone interviews with participants at the end of the season. Of twenty-one interviewed, nineteen wanted to continue for another year if they could pay the same amount. These are some of the comments:

"This was a fabulous program! Thanks so much for letting me take part. It has given me lots of new cooking ideas and I can't tell you enough that I so enjoyed the fresh food all summer. I was sorry to see it end. Thank you."

"I enjoyed the farmers' letters and tried recipes I never would have before."

"I really appreciated the program. I feel healthier, and no one in my family got sick this winter. I think it is because they ate so many vegetables this summer."

"There is a need for the program because a lot of families and kids never get fresh food; if they do, it is usually at Thanksgiving, and the produce only consists of root vegetables because of the time of year."

"My kid actually eats more fresh vegetables than when they are canned."

"Having this food for our teenage clients meant that they were not going to McDonalds or the Kerry Quick Stop as much."

"I would like to have our kids participate in the Farm Share Program again and to help them to get to know the CSA community more."

"I am unable to purchase fresh produce now because it costs too much in the winter. My finances are tight. I just saved $40 using coupons, but there aren't any coupons for fresh food. I sometimes purchase dented canned vegetables at a discount store."

COMMUNITY FOOD SECURITY

As the government safety net weakens, and awareness of the meaning of community food security spreads, projects have been springing to life around the country to help people meet their own food needs. Linking with community supported agriculture has seemed a natural idea. The Coalition for Community Food Security defines food security as "all persons in a community having access to culturally acceptable, nutritionally adequate food through local non-emergency sources at all times." The Coalition encourages the establishment of local and regional food system councils to do long-range planning, and the creation of community-based networks and coalitions to strategize and implement multi-faceted programs. Most community food security projects are sponsored by nonprofit organizations, which are seeking innovative ways to solve the complex and interrelated problems of hunger and poverty in the current food system. None of these projects is more than ten years old, and many are just getting underway as I write. Although a few of them involve independent farmers as advisers or as sources for some of the food, most like the Food Bank of Western Massachusetts, have established new farms. Like community supported agriculture, community food security projects are experiments with new social and organizational forms, working with the slimmest of resources and combining inherently fragile human materials.

One of the earliest of these projects is the Homeless Garden in Santa Cruz, California. I was fortunate to have the chance to tour the garden in 1992 when it was in its second year at its original site, in the middle of a neighborhood of fairly expensive homes. Read Jered Lawson's vivid account of the Homeless Garden, written for Rain Magazine in 1993, in the sidebar on pp. 198–99.

When Jered wrote his article on the Garden, its situation was precarious: The city owned the land and was planning to subdivide it to sell as building lots. When I spoke to current director Rick Gladstone in March 1998, the Garden had finally been forced to move to another two-acre piece of land, but was negotiating with the city for five acres in a greenbelt planned for the edge of town. According to Rick, their fate depended on the outcome of a debate raging between "deep" ecologists, who want to keep the land as a nature preserve, and "social" ecologists who favor the Garden. Rick was focused on fundraising, feeling under pressure to expand sales and reduce grants, but frustrated in doing that by the lack of additional land. He remarked that it seemed "incredible that we are expected to support the entire project as a farm business." A letter from Rick in September 1998 announced that the social ecologists had won, and the Garden would be moving to its new site at Pogonip. Jered's conclusion is as true today as it was in 1993: "The Garden demonstrates that ecologically sound, socially just, and economically viable projects are possible. What's needed now is the motivation, determination, and commitment of individuals who recognize the potential in people, and the land, to heal, take root, and grow. Michael Walla of the Garden says 'the Garden is showing we're people with pride, people willing to struggle. . . . We don't need someone who will carry us. We need someone who's willing to help get us on our feet.'"

Similarly training and employing homeless people, Carrie Little leads the Tahoma Food System project in Tacoma, Washington. Funded by grants for the first three years, the project is struggling to be self-sustaining with the support of sixty CSA shares priced at $450 each, sales of dried flowers, vegetables, and herbs, and garden tours. For 1998, Carrie's salary will come from a USDA Community Food Project grant that she hopes will enable them to provide training in organic farming and marketing to Southeast Asian immigrants who grew food in the old country. The Tahoma Food System works closely with Guadelupe House, a Catholic Workers project. Carrie, who discovered the joys of gardening in seventh grade biology, and six part-time trainees grow the food for the CSA shares on nine urban garden sites within a ten-block area, a total of 4½ acres under cultivation. Each of the six trainees oversees a separate site. The majority of the CSA members are white and middle-class, but Carrie says there is a great mix of ethnically diverse people as well. As if this were not enough to do, Carrie is trying to get a second project going in the nearby city of Edgewood, which has purchased a 10-acre farm.

Most community food security projects are sponsored by nonprofit organizations, which are seeking innovative ways to solve the complex and interrelated problems of hunger and poverty in the current food system.

THE HARTFORD FOOD SYSTEM'S HOLCOMB FARM CSA

The Hartford Food System (HFS) in Hartford, Connecticut, is one of the oldest, most consistently inventive, and productive of the community food security nonprofit organizations in the country, providing a model of the systemic approaches advocated by the Coalition for Food Security, of which it is a founding member. Since 1978, under the leadership of Mark Winne, HFS has worked to plan, develop, and operate local solutions for the City of Hartford's food problems. To help save area farmland by improving the earnings of local farmers, while increasing the supply of fresh, nutritious food for city residents, HFS rallied Hartford city agencies and community organizations to establish the Downtown Farmers' Market, the first of forty-eight farmers' markets in the state. To enable more low-income people to shop at these markets and increase sales for local farms, HFS worked with the Women, Infants and Children (WIC) program, the Connecticut Department of Agriculture, and other agencies to develop the Connecticut Farmers Market Nutrition Program, which annually provides over fifty thousand WIC recipients with $396,000 in special vouchers that they can spend only to purchase fresh produce from area farmers' markets. HFS has also worked with agencies for the elderly to fund $90,000 worth of these coupons for six thousand senior citizens.

The Homeless Garden

by Jered Lawson

A year and a half ago, Bill Tracey stood on the corner of Chestnut and Mission clutching a piece of cardboard that read "Homeless and Hungry: Will Work for Food." Now Bill works with a group of homeless people who take these words literally, growing food not only for themselves, but for the surrounding community as well. In just two years, over forty homeless people, a committed staff, and countless volunteers have turned a 2½ acre vacant urban lot in Santa Cruz, California, into a thriving organic garden.

After the gardeners take their portion of the harvest, much of the food goes to community members, or "shareholders," who support the garden financially. A percentage of the produce is sold to local stores, restaurants, and folks at the farmers' market. The rest is donated to homeless shelters and free-meal programs. Bill says "other homeless projects can give you files, reports, and statistics, but we can give you a flat of strawberries."

The garden offers diverse flora with a mixed crew of gardeners. There's Peter, a homeless trainee; Darrie, a mother of two; Paddy, a volunteer handyman and gift-giver to the garden; and Phyllis, a vivacious 82-year-old who asserts, "I don't have to die to get to heaven . . . this place is heaven on Earth." Both Mac, a humorous and stately homeless man, and Mike a "practical idealist" university intern, work with groups of children in the garden. According to Lynne Basehore, the project director, "The garden has been useful to those who simply need to witness life's abundance. Most of all, it has been a renewal for long-term jobless and homeless citizens of the community."

With over two thousand homeless people in Santa Cruz County, it's no wonder there's a waiting list for the fifteen paid positions available at the Garden. When a position does open, prospective employees volunteer a short while to see if they are truly interested in the work. If so, they begin at a minimum of twelve hours a week and attend the weekly meeting. Workers are familiarized with procedures of the garden, and then choose an area for in-depth training. For Skooter it was compost, for Octaciano, the greenhouse. Jane Freedman, the garden director, trains the gardeners in bed preparation, composting, cultivating, planting, harvesting, and selling produce in the farmers' markets.

The weekly meetings provide group members with an opportunity to air concerns, make collective decisions, and work through any pressing problems. A rules committee—made up of five of the homeless workers and two of the staff—compiles and presents a list of rules that are then agreed upon by the larger group. Developing and enforcing their own rules gives the workers a voice in decision-making that they were generally denied elsewhere. Some of the rules: "When scheduled for work, do not come high, drunk, or hungover: if you do, you will be sent away immediately and the consequence is suspended paid work until nine hours of volunteer work are completed." "No sleeping at the Garden. Anyone caught camping is kicked off the project."

The Garden's pay, \$5–6 an hour for twelve hours a week, may not be a living wage. But Darrie Ganzhorn, a garden employee, feels the money is "only one piece of the puzzle. It's part of the network needed to get one's life together." Another gardener says, "the work has grounded me. It's stabilized me to where I can actually go out and enroll in school. Otherwise I'd be too scattered. You know, hustling to get this or that. Since I've worked here, I've moved to a safe place to sleep at night."

History

Paul Lee, an internationally renowned herbalist, former UCSC Professor of Philosophy, and long-time advocate for the homeless, inspired Lynne

Basehore and Adam Silverstein in May of 1990 "to transform the vacant lot into a healing, productive garden." Paul, after receiving a donation of herb plants from a store in Carpinteria, California, knew that "if we had a couple thousand plants on hand, we would have to get them in the ground; hence, the Homeless Garden Project!"

Lynne began recruiting homeless workers from the shelter to come and work for a few hours here and there, getting the herbs in the ground. Since the herbs needed watering, and since Adam had experience in irrigation systems, he too became a part of the crew. Jane Freedman became director in November 1991. In reference to the "horticultural therapy" aspect of the garden, Jane once joked, "We may not have any couches, but we certainly have a lot of beds."

The Garden uses Alan Chadwick's French in tensive/Biodynamic, raised-bed method of gardening. Local restaurants, cafés, horse stables, landscapers, and neighbors give [organic materials] to an innovative composting system.

Money for salaries and wages comes from a variety of sources. One-third of the budget is covered by the CSA, as well as through the sales of produce and flowers at the local farmers' markets, restaurants, and natural foods stores. Funds are also raised through special events, grant and letter writing, awards, and direct campaigning. The project was selected by Visa card holders of the local Santa Cruz Community Credit Union to receive 5 percent of the money generated from the use of their cards. The New Leaf Community Market began a unique system of fundraising, issuing 5¢ "enviro-tokens" to shoppers returning paper bags. The tokens are given to the nonprofit organization of their choice. So far the Garden has been the community's favorite, generating more than five thousand tokens in three months. And finally, the Garden receives subsidies from the local Job Training Partnership Act (JTPA), the American Association of Retired People, and the Veterans Affairs Job Training Program.

The gardeners also receive benefits from the community. One vegetarian restaurant gives the project free meal tickets in exchange for produce. A local laundromat, "Ultra-mat," provides the gardeners with a monthly allotment of "Ultra-bucks" to use for washing their clothes. Some of the gardeners have buddied up with volunteers who assist with basic needs; from a bed roll for the night, to a job or housing opportunity. The Garden is also a magnet for contributions: clothes, a computer, and even a couple of trucks.

In the fall of 1991, the Garden adopted the CSA model. Shareholder Steven Beedle says "the CSA has meant guaranteed access to the freshest organic produce at a great price, supporting the much-needed assistance to homeless people, and having a say in all issues that are confronting the Garden. There's a sense of involvement, with the people doing great work and benefitting in the process."

Source: Rain Magazine 14, no. 3 (Spring 1993): 2–7.

Since he helped initiate this program in 1987 in Connecticut, Mark Winne has facilitated the spread of the Farmers Market Nutrition Program all over the country with joint funding from USDA and the states. HFS Program Director Elizabeth Wheeler is working with the public schools in Hartford on Project Farm Fresh Start to find ways to incorporate locally grown organic foods into the school lunch program. HFS has also helped bring new supermarkets to inner-city neighborhoods, and has initiated city- and state-wide food policy councils to plan for greater food security. Mark Winne was a guiding member of the Community Food Security leadership, which achieved passage of the Community Food Projects Program as part of the 1996 Farm Bill, providing $2.5 million a year in competitive grant money to projects such as the Tahoma Food System.

In its efforts to address the critical need for sources of fresh, nutritious produce in Hartford, HFS has worked to create links between growers and low-income urban consumers, and for many years sought the opportunity to establish a CSA. In 1993, the Friends of the Holcomb Farm, a nonprofit Granby organization, invited HFS to participate in its plans for the Holcomb Farm Estate, which was bequeathed to the Town of Granby by Tudor and Laura Holcomb. In public hearings, and not without some dissenting voices, the people of Granby approved two main goals for the farm: to maintain the farm for agricultural use, and to make the farm available to the wider community, with special outreach to Hartford residents. The plan was aimed to create a CSA that would cultivate sixteen acres of fruits and vegetables, half of which would go to low-income Hartford residents and community organizations. The lease signed by HFS also stipulates that for every five acres cultivated, the farm will provide two shares to Granby social services.

The HFS staff runs the CSA operation, handles all the money, and does all the hiring and firing of staff for the farm. During these startup years, HFS fundraising has covered most of the expenses of purchasing equipment and improving the land. Elizabeth Wheeler, who is in charge of this project for HFS, says that their plan is to sell enough shares to cover all the operating expenses. Shares accounted for 52 percent of the operating budget in 1997, the third year. Donations of all kinds from local farmers and members of the Granby community have helped reduce cash outlays. The farm staff call Elizabeth the "queen of donations." The community groups pay about 25 percent of food costs: HFS makes up the difference with grants and donations, and expects to continue so that the CSA need not bear that added burden.

The farm staff consists of a farm manager, an assistant manager, and interns; Vista Volunteers facilitate the participation of the Hartford groups. Like most other farms run by non-farming organizations, the Holcomb Farm has had a steady turnover of farm managers. Mike Raymond, the fourth manager in five years, took over from Alexandra Smith in 1998. The organizing and support provided by Elizabeth is clearly crucial to the smooth functioning of the CSA. She assembles annual reports full of detail that would be invaluable to anyone considering a similar project. In 1997, a staff member compared the value of the family shares with a comparable amount of food from a large supermarket selling conventionally grown produce and a natural foods store selling organic produce. The same amount of produce would have been worth $337 at the supermarket and $535 at the natural foods store. Organizational members saved $5,000. In 1996, Holcomb Farm charged members $1.05 per pound, compared to $1.01 per pound paid by members of Food Bank Farm of Western Massachusetts.

Holcomb Farm CSA is committed to collaborating with area farmers, who provide crops such as organic corn and low-spray apples. In 1997, the CSA offered fruit shares of apples and pears from Ginny Wutka's Lost Acres Orchard and corn from George Hall of Simsbury, who has also started his own CSA. Sixty-four households and two organizations purchased fruit shares.

Elizabeth Wheeler related some of Holcomb Farm's history at the Northeast CSA Conference in November 1997:

> There was some emotion broiled up among the citizens of Granby by this new approach. To do a subscription farm in a town that has an existing farm community was quite a radical idea. I did get a few phone calls from people wanting to know when the buses with those people were coming and whether their social security numbers would be written down, and a few calls from irate farmers claiming we were unfairly competing with them because we were going to steal customers.
>
> In Hartford, it was my job to find the folks we were trying to reach. Hartford is the eighth-poorest city of its size in the U.S. To give you an indicator, 80 percent of the twenty-three thousand children who go to school in Hartford are on free or reduced-price lunch. We have sections of the city where there is 40 to 50 percent unemployment and some fairly dire circumstances. Obviously, for a low-income individual in Hartford, who has not set foot outside the city in some time and doesn't have a car, the CSA was not going to work. The best approach was to go to community organizations that were already working in the city on economic development, housing, and the like, and recruit them to take on responsibility for distributing the food.
>
> In terms of connecting diverse people, it's a real slow process. Folks who aren't used to being with one another are not necessarily going to *want* to be with one another, and it's taken a little bit of social engineering to bring them together. We have arranged some special events. When we bought our new tractor, we invited the black clergy from the city and some of the white suburban clergy to bless it, a wonderful symbolic event. Many little connections and bonds have been made. It's exciting to see this happening through the leveling connection of food.

The CSA project has an advisory group with four people who are representatives of households and four from the city organizations. The head farmer also participates. In 1996, the advisory group worked with HFS staff and a consultant to develop a three-year strategic plan for the farm to guide budgetary and policy decisions. The group established the following goals:

- To grow the quality and variety of produce desired by CSA members;
- To foster responsible stewardship of the land by using exclusively organic methods;
- To restore the line between people and agriculture by encouraging involvement in the farm;
- To address the problem of hunger and malnutrition in Hartford;
- To give city residents the opportunity to participate in activities at the Holcomb Farm;
- To create a model of regional cooperation and exchange for urban and suburban communities;
- To operate a self-supporting farm by the year 2000.

To get more detail on how the community groups distribute their bulk shares, I called three of them directly. Staff members of Family Life Education, which serves over two hundred families on welfare, bring bags of food to the families who are most in need. A few times, they have been able to bring families to the farm to help. Terry Finn of the YMCA Youth Shelter says that they got involved with the farm in 1994 because it was a "good deal, a lot of vegetables at a good price." The vegetables inspired the shelter to start a farm stand, where the teenage girls served by the shelter get the experience of selling, and raise money for other activities. Weekly they sold $40 to $50 worth of vegetables as well as salsas and other products they made themselves. For the girls who really get involved and do well, Terry says, it helps build self-esteem. As to the weekly farm work, Terry reports that "some girls love it, some hate it—there's not much middle ground." O.N.E. C.H.A.N.E., Inc., a nonprofit devoted to rebuilding Hartford's North End through community organizing, ownership housing development, job training, and youth programs, gives out the CSA shares through its seventeen block clubs. Their youth teams of fourteen- to seventeen-year-olds have helped plant, weed, and harvest crops at Holcomb Farm. "No one can ever imagine the impact this program had on the youths we serve," said Gerald Fullwood, a community organizer. "By the end of the program you could actually see the difference made, and it gives you a good feeling. Not only did the kids get a great experience working on a farm, but very needy families in our communities got a lot of fresh vegetables for free."

TURNING FARM WORKERS INTO FARMERS

One of the most unusual and exciting CSA-related training programs is taking place at the Rural Development Center (RDC) in Salinas, California. Founded by José Montenegro in 1985, the RDC provides agricultural training, education, and marketing experience to farm workers to give them the skills they need to run their own farms. Over 80 percent of their graduates go on to become independent farmers. Upon completion of a five-month course in agricultural production and farm management, participants can use RDC land, water, equipment, and continuing technical support for up to three years. A 1997 Community Food Project grant will enable the RDC to expand its training program for farmers, community gardeners, and schoolchildren, to set up self-supporting food distribution and marketing, and to establish a Public Education and Policy Council to promote food security efforts. Staff member Luis Sierra told me that, before 1991, the 112 acres of RDC land was managed using conventional farming methods. That year, one of the participants grew one acre of zucchini organically. The productivity and earnings were so impressive that by 1998, 86 of the acres had been converted to organic methods. Luis gives the students technical assistance with marketing their crops, most of which they sell to brokers. Having interned at both Full Belly and Live Power Farms, Luis has been encouraging the students to try out CSA.

Maria Iñes Catalán, one of the first participants to try organic methods, marketed her organic vegetables for three years through brokers. The RDC allows her to continue using their land because she is an innovator. In 1996, with the help of Luis, she started a CSA. She grows about forty-five different crops, with twelve available throughout the season. An ad in a free local newspaper and recruiting by members of the RDC advisory board helped sign up her first sharers. Her CSA does not yet have a strong group of consumers committed for the whole season. In 1997, the number of shares ranged from twenty-two to forty-two, with many only staying for six or eight weeks. Maria's goal for 1998 is to get forty for the whole season. The members, who are mainly white and middle-class, live in the towns of Salinas, Carmel, and Monterey, where Maria has drop-off centers.

Maria does not speak English, an obstacle to organizing consumers. Sometimes the RDC helps her. Other times, she relies on her four children, ages eight to eighteen, to translate. (My son Andy interviewed her for me since he speaks fluent Spanish.) For 1998, a woman will help with communications in exchange for a free share. Maria has produced a CSA brochure in Spanish in the hopes of recruiting in the Hispanic community. Finding Latino members has proved difficult because the CSA is in the middle of a major vegetable crop region. Many Latinos obtain free food through connections to the vegetable industry, or grow their own. Maria has found that she can attract Latinos by growing varieties, such as chiles, that are not available commercially in the area. Her crop list would be helpful for CSAs elsewhere that seek to serve Hispanic communities (contact the RDC, listed in CSA Resources).

Two of the other RDC student farmers are initiating CSAs in 1998. Luis sent me a copy of Valdemar Alonso's CSA brochure, which offers a typical crop list with a price of $14 to $16 a week, depending on the payment plan. Members can make a commitment as short as one month. Valdemar will also try to market a box for $10 filled with basic Latino vegetables, such as potatoes, onions, garlic, Mexican corn, jicama, cilantro, tomatoes, and sweet and hot peppers. So far, Luis reports, neither the CSA model nor farmers' markets have worked very well in the Hispanic community near the RDC. One graduate was successful selling vegetables door-to-door out of a truck. Luis is determined to keep trying different direct sales angles until he finds some that work. He is exploring the possibility of institutional CSAs, connecting his RDC growers with churches, WIC, or Head Start programs with active parent groups, or selling to the central kitchen that supplies the local schools.

COMMUNITY FOOD PROJECTS

The availability of grant funding from the Sustainable Agriculture Research and Education and Community Food Projects programs has stimulated the expansion of a number of projects that combine training in food production for low-income people, the use of university land and student energies, and CSAs. To receive Community Food grant money, projects must:

- meet the food needs of low-income people;
- increase the self-reliance of communities in providing for their own food needs; and
- promote comprehensive responses to local food, farm, and nutrition issues.

The outstanding qualities that unite these projects, however, are the enthusiasm, high energy, and determination of the organizers, and the unusual mixture of social service, farming, church, Extension,

undeniable as a dog's smile

in just an instant
the belief was made manifest.
the early evening
light was aslant and full
of flying creatures,
myriad motes and swirling
insect constellations.
swallows pirouetted above the melon
vines in an ecstasy of bug catching
while a phoebe sat atop
the fork handle and watched,
flicked her tail then hovered
and nabbed a meal.
black cat stalked amidst foliage
where soldier bugs sucked away at
immature potato beetles, fat and
orange and
softly similar to the ladybugs walking by.
in that instant,
as sudden and clear as the light
in a loved one's eyes,
the belief in our interconnectedness
was visible.
and with it, undeniable
as a dog's smile, comes the need
to honor the balances
and to nurture the creation
instead of
randomly
extinguishing its parts.

—Sherrie Mickel
1992

university, and community organizations that have come together around local food. No two of these projects are the same. It will be fascinating to revisit them in a few years to see what has worked and what lessons they have learned.

IN IOWA, THE FIELD to Family Community Food Project, with funding from the Kellogg Foundation, the Leopold Center for Sustainable Agriculture, and the Community Food Projects Program, will reach out to low-income families through church groups and social services to involve them in the rapidly increasing network of CSAs, nutrition education, hands-on farming and garden work, and leadership training. Members of Magic Beanstalk CSA initiated the Field to Family Project to foster more self-reliant local food production and to help lower-income families gain access to wholesome food. Robert Karp, co-director of Field to Family, explains: "Instead of just being given a food handout, low-income families are invited to participate in a community process that supports local farmers." Robert developed the idea of using CSA as a focus for delivering social services to low-income families. Grants from national churches subsidized seventeen shares for low-income families in 1997 and twenty-five shares in 1998. Magic Beanstalk donated over 3,500 pounds of food to local food pantries. (See chapter 19 to read about the Magic Beanstalk CSA in Ames, Iowa.) In 1998, Field to Family organized well-attended community meals, started a new downtown farmers' market, held monthly cooking classes for adults and children, sponsored Iowa-grown meals with several conference centers, and began planning a food-processing microenterprise.

IN MISSOULA, MONTANA, on two acres of land donated by the University of Montana and two acres of county land, Josh Slotnick, his students, and volunteers grow food for eighty shares for the Garden City Harvest CSA and give away as much to WIC families and the local food bank. The students do six hours of farm work per week for Josh's PEAS course, the Program for Ecological Agriculture and Society. Local people can "Volunteer for Veggies" and earn a whole weekly share by working eight hours, or a half share for four hours. Welfare recipients can fulfill their twenty hours a week of community services by participating either in the CSA or the five acres of urban gardens. An Ameri-corps Volunteer coordinated the seven hundred volunteers in 1997. Garden City organizes gleaning and encourages gardeners to "grow a row" for the food bank. To make Garden City financially self-sufficient, the project sells the shares, offering one share size but three fee choices: "living lightly"—$180, "middle of the road"—$250, and "have enough to share"—$320. They are also hiring a farmer to grow echinacea on another five acres of county-owned land; Trout Lake Farm will buy and process the medicinal herb. Mary Pittaway, a public health nutritionist for twenty-five years, oversees the entire Garden City project for Missoula Nutrition Resources. Her real job is WIC director, but she finds the food production project the most fulfilling work of her career and the most effective approach to nutrition education she has seen. Their slogan is, "Together we are learning self-reliance, and the medium is food."

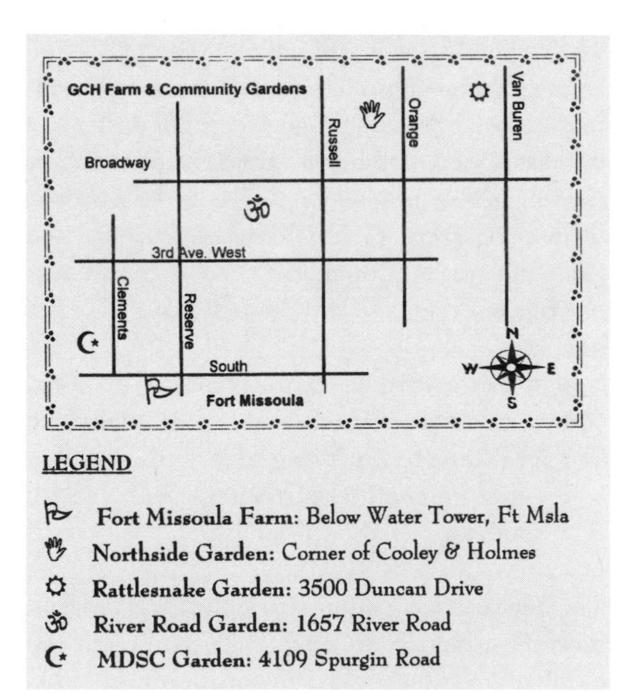

Garden City Harvest CSA, Missoula, Montana

SMOKEY HOUSE CENTER in Danby, Vermont, employs local teenagers, who learn work skills on the job in carpentry, forestry, animal husbandry, and gardening. The primary markets for the center are a thirty-member CSA, a local farm market, and area restaurants. Besides vegetables, the teenagers, under the tutelage of Ray Pratt and Theresa Hoffman, produce maple syrup, hardwood charcoal, Christmas trees, blueberries, pork, beef, lamb, and wool. The nonprofit's main product, however, is the training for the teenagers.

ISLES, INC., IN TRENTON, NEW JERSEY, is starting a farm on Mercer County College land to grow food for soup kitchens and food banks, while training people on general assistance in horticultural skills. Isles is an eighteen-year-old community development organization with projects in affordable housing, vocational and leadership training, environmental education, and public health in the largely Hispanic and African-American inner city. The farm project's enthusiastic coordinator is Ron Friedman, who ran a nursery business before going to work for Isles five years ago. As the project was getting underway during the winter of 1998, Ron's biggest concern was getting willing participants. City social services screens applicants and administers drug and medical tests, but Ron has the final say. His plan for financial self-sufficiency by the end of the three-year grant period is based on selling shares, setting up inner-city farm stands, supplying flowers to the strong local flower market, and producing annual and perennial plants on contract to local nurseries. With five acres available, Ron is beginning production on one acre, plowed and disced by a local farmer, and trying for ten shares the first season.

IN PHILADELPHIA AND CHESTER, Pennsylvania, Sea Change combines a CSA on urban garden sites with training for teenagers. Concerned with environmental justice, Sea Change sponsors a series of imaginative programs aimed at low-income people: a demonstration project on aquaculture, a charter school, a Cyber café, and a bicycle recycling project. Since 1995, Sea Change has been renting garden plots to families and offering technical assistance in using organic methods. On Saturday mornings, they hold a farmers' market. In 1997, with funding from SARE, they launched a CSA with five full paying shares and three sponsorship shares divided among ten families who could not afford the share price. They plan to increase the shares to twenty-five in 1998 and add sales to restaurants. A group of forty teenagers does the farm work under the guidance of an experienced farm manager. The director of Sea Change, Rosalyn Johnson, says that they do not expect the farm to pay for itself for a few more years. The pressures of welfare reform, Rosalyn explains, led to their focus on learning how to grow clean, healthy food in a small space. The food at neighborhood stores is often low in quality, even rotten. Sea Change wants to integrate nutrition, health issues, and food production. "If the going gets rough," Rosalyn says, "it is important for kids to know how to grow food for themselves; they can plant some seeds and feed their families." In July 1998, the city redevelopment agency ordered Sea Change to vacate their garden site by October 1, 1998 to make way for a parking lot. Protests from food system advocates throughout the northeast persuaded the agency to allow Sea Change to keep the land until the end of the growing season, and to offer another site for the next year.

AS SOMEONE WHO HAS FARMED on a small scale in the quiet of the countryside, I find the projects I have just described incredibly complex organizationally, admirable in motivation, and perilously ambitious in scope. Every one of these projects deserves to have its full story told. I hope their daily struggles and moments of victory will be carefully documented. Their ultimate success will depend on many fragile elements coming together in just the right balance. If Community Supported Agriculture does nothing more than to help redress some of the injustices in our present food system, it will have earned its place in history.

AGRICULTURE-SUPPORTED COMMUNITIES 21

God said: "I have provided all kinds of grain and all kinds of fruit for you to eat."

GENESIS 1:29

IF IT IS NOT ALREADY OBVIOUS that the CSA concept has inspired many different kinds of projects, in this chapter we offer four short essays on CSAs initiated by faith groups, on city-owned land, on a college campus, and in Japan, where it all started. While farmers have created most of the existing CSAs in North America, reaching out to local consumers for financial and moral support, a variety of communities are seeking to find or train their own farmers. Community Supported Agriculture—Agriculture Supported Community; as Robyn Van En noted at the beginning, the mutual relationship is what is important. At the November 1998, conference of the International Federation of Agriculture Movements (IFOAM) in Argentina, I had the pleasure to meet Shinji Hashimoto, a teikei farmer from the Kobe area in Japan. Back in 1993, a group of Kobe teikei members and farmers had visited a NOFA summer conference. When the earthquake flattened Kobe, I worried about their fate. Shinji showed me a photo that answered my question: his blue pickup truck loaded with tanks of water and vegetables parked next to the remains of some houses. The teikei farmers provided drinking water and free food to their consumers—the ultimate in agriculture supported community!

FAITH COMMUNITIES ON THE LAND

*by Patricia Mannix**

In the late 1970s and early 1980s, people working on issues of peace and justice, especially in the area of world hunger, became painfully aware of the connections between land abuse due to industrialization, war, and greed, and the growth of poverty, disease, and hunger across the planet. An Earth-based spirituality began to

* Patricia Mannix, long-time social activist and organizer, wife, mother, and grandmother, is a founding member of GVOCSA.

emerge with a New Cosmology of interconnectedness. The writings of Thomas Berry, C. P., a Passionist priest from the hill country of North Carolina, now a cultural historian, served as vital chapters in the burgeoning ecological apocalypse.

ONE PLACE THE NEW COSMOLOGY found fertile ground was in Roman Catholic religious orders. Early on, Sister Miriam Therese MacGillis, a Dominican sister from Blairstown, New Jersey, became a convert. In the anthology *Earth & Spirit,* she warns us, "Unless our human species can open itself to the contemplation of food as a holy mystery through which we eat ourselves into existence, then the meaning of existence will continue to elude us" (p. 163). In 1978, when her order inherited the 140-acre Red Cat Farm, Sister envisioned it as a place to welcome people of good will and teach them about the Earth. Acceptance of her proposal in 1980 led to the birth of Genesis Farm, which is now known worldwide as a "learning center for re-inhabiting the Earth" (p.161). In 1987, Genesis Farm attracted a Biodynamic grower who had a desire to develop a market farm. However, they soon realized that there was no way they could grow food, produce a just wage, and still market it at a competitive price. After seeing Robyn Van En's video "It's Not Just about Vegetables," they knew that Community Supported Agriculture was the answer. Thus began the venture of Genesis Farm into "sacred agriculture."

In keeping with sustainable land use, the folks at Genesis have determined that the 13 acres of garden plus the orchards will produce only enough food for one hundred shares. Members tend to be people who were already ecologically and spiritually aware. Word of mouth is usually sufficient to recruit the needed members. The relatively high price of the shares ($1,340 for a full yearly share) reflects the passion for justice at Genesis Farm. The gardeners are paid good salaries with benefits, which include housing allowances and money toward retirement. In exchange, members receive an amazing array of Biodynamically grown produce.

Sister MacGillis said a wonderful part of the whole venture is seeing the people begin to care about each other, and watching the children grow from babies to teens. While admitting that she does feel tension each spring, wondering if there will be enough people joining, she still wanted mostly to talk of "the magic of the garden and the food itself." Sister says it is a very special moment when it is apparent that a member feels the farm is home. When I pressed her to name the very best part of the community supported garden, she lovingly declared, "Just all of it!"

SISTER JANIS YAEKEL, A.S.C., is the director of Earthworks, an ecological education facility loosely fashioned after Genesis Farm. Earthworks was started through the vision of a group of sisters called the Poor Handmaids of Christ, who own fifteen acres of beautiful land on the shores of a small lake in Plymouth, Indiana. On three acres of pastureland, the sisters began a community supported garden as an alternative economic and agricultural program to foster education about care for the Earth, and to build community around food. Membership reached eighty shares in 1996, but not without many organizational and staffing problems. "The Garden" project is taking a break for two years while reconstruction goes on to increase the accommodations at the Earthworks education center. Sister Yaekel says there will have to be much stronger member participation when the Garden does restart in 1999. Despite a rocky beginning, Earthworks is becoming more stable and is looking forward to the future with the Garden. Sister Yaekel says they have adopted the positive attitude displayed in the film *Field of Dreams,* with a little twist: "If you plant it, it will grow."

IN A DISPLAY OF SERENDIPITY often found in things of the spirit, Sister Gail Worcelo, C.P., and Sister Jeanne Clark, O.P., who had not met at the time, were both led to similar names for their projects. Sister Worcelo is the director of Homecomings, a center for ecology and contemplation in Clarks Summit, Pennsylvania, while Sister Clark lives out her ministry at Homecoming, a community dedicated to bioregionalism and sustainable living in Amityville, on Long Island, New York. The two groups are connected only in vision.

Sister Worcelo started a Community Supported Garden (CSG) at St. Gabriel's Monastery in 1992 as part of Homecomings, a ministry of the Passionist Sisters. The sisters also run a retreat house offering spiritual direction, and bake the breads used in the Roman Catholic celebration of the Eucharist. The sisters wanted to share their land with the local community, to give them the opportunity to eat organic food, and to realize that how we eat determines how the world is used. This small venture had only twelve members in the 1997 season. The members are ordinary people from the surrounding community, many of them not previously aware of ecological issues. Sister Worcelo feels that the failure to develop a consistent operating routine over the six years has hindered membership growth.

Across the country, many different church groups are finding nourishment of both body and soul in sharing the land.

The share costs only $60, kept low to encourage all people to join. Persons who cannot fulfill the work requirement of two hours per week must pay an extra $40 to $50 a season for their shares. In a true spirit of justice and community, the share has no set amount of produce: members are allowed to take what they want and need for their families from whatever is picked on a given day.

Sister Worcelo feels the best part of the CSG is the opportunity it gives the local folks to form community among themselves. She also appreciates the opportunity the garden presents the monastery to extend into the greater community. While the lack of a set operating procedure has been difficult, Sister also sees it as an advantage, since it more truly represents the organic nature of life, growing and ebbing with the conditions, requiring different things at different times. Just letting the organization grow as it will becomes a plus and a minus at the same time.

THE 1998 SEASON WILL BE the first year shareholders partake of the riches of Sophia Garden at Queen of the Rosary. This community supported garden grew out of Sister Clark's project, Homecoming, and is part of the ministry of the Dominican sisters at Amityville. Homecoming started in 1994 as the plan of a group of people (not all nuns) to purchase some land where they could live communally and teach people how to "come home to Long Island" through an understanding of bioregionalism and sustainable living. Long Island is a very consumer-driven, stressful place, and land proved to be too expensive. Still holding to their dream, the group leased a half acre of land owned by the Dominican order, and Homecoming was born. At present, it is still a part of the order, but is moving toward a more independent status.

In 1997, the project planted cover crops on one-quarter acre, and vegetables, pear and apple trees, and flowers on another quarter acre. The first year, all produce went to the Motherhouse. The garden was a success, and so, in 1998, Homecoming sold the first shares.

The one hired gardener is assisted by a variety of youngsters, many from the surrounding African-American community, who come from the "Weeds and Seeds" program in Amityville, and are paid minimum wage for four hours of work per day during July and August. Weeds and Seeds generally has nothing to do with agriculture, but is a military-type venture with at-risk youngsters designed to "weed out" the bad influences in their lives and plant "good seeds" instead. Over the years, the sisters have not had the presence in this community that they might have had, and Sister Clark sees the garden as one way to reach out.

ACROSS THE COUNTRY, many different church groups, embracing several different denominations, are finding nourishment of both body and soul in sharing the land. On Common Harvest Farm, in Osceola, Wisconsin, Margaret Pennings and Dan Guenthner and their children are proving it is possible to practice simple living in an urban setting while joining with their intentional spiritual community to put their faith into action. Dan interned on a farm and belongs to a Lutheran faith group called the Community at St. Martin's. Believing that food is important in community, and that urban people are often cut off from good food, he and Margaret decided to form a CSA and rented some land in 1989. In 1998 a dream came true when they were able to buy their own farm. Members put up $60,000 in no-interest loans and gifts that are tax-deductible, thanks to a conservation easement on the farm.

The CSA has grown steadily, and in 1997 had 171 shares. Dan says this is probably larger than it should be. Much of the membership comes from faith communities: Baptists, Unitarians, the United Church of Christ, Lutherans, and Roman Catholics. Dan, having been in seminary once, finds working with congregations very comfortable. Grants from the Evangelical Lutheran Church of America have enabled the CSA to grant share scholarships, and sometimes they just give food to those who need it. They are very forgiving of folks who fall behind in their payments. Even those who have taken as long as two or three years to catch up are never cut off or prevented from signing up for the next season.

The Community at St. Martin's is committed to proving that a farm, built on a strong spiritual foundation, can last "forever." They would probably agree with Sister MacGillis when she says, "When we understand that food is not a metaphor for spiritual nourishment, but is itself spiritual, then we eat food with a spiritual attitude and taste and are nourished by the Divine directly" (*Earth & Spirit,* p.163). And all can join with the psalmist and declare, "O Lord, my God, how great you are! You make grass grow for the cattle and plants for humans to use so that they can grow their crops and produce wine to make them happy, olive oil to make them cheerful, and bread to give them strength" (Psalms 104:1,14–15).

THE INTERVALE: COMMUNITY-OWNED FARMS

*by Beth Holtzman**

Being a member of the Intervale Community Farm is at the heart of who Dia Davis is. "I live in the city, but I'm growing up on a farm," the freckled eight-year-old cheerfully says before trotting off to help sow vegetable seeds.

During the growing season, Dia and her mother, Bonnie Acker, volunteer three or more days a week at the farm, located within Vermont's largest city. While the Acker-Davis family spends far more time in the fields than most of ICF's 350 members, it's clear that the spirit of inclusion and cooperation has helped this consumer-driven CSA flourish. Acker exclaims,

> It's been the greatest adventure of our lives because it's all cooperative. Every day there's a crisis or problems that need to be dealt with, and there's always success. . . . And when we leave, we always have the sense that we've been truly helpful and truly appreciated. There aren't many places in this culture where people can participate so fully in a cooperative venture, especially children.

The Intervale Community Farm (ICF) is one of several examples of farms on town-owned land in New England, along with the Natick Community Farm in Natick, Land's Sake in Weston, and Codman Community Farms in Lincoln, all towns in Massachusetts. In 1998, the Intervale Community Farm was the largest, in terms of membership, and one of the oldest CSAs in Vermont. Since it began in 1989, the farm has fine-tuned the details of share size, crop mix, member work requirements, staffing, and salaries to develop an operation that is both financially stable and highly attractive to Burlington-area residents.

* Beth Holtzman is the Communications Specialist for the Northeast Sustainable Agriculture Research and Education Program and a member of a Vermont CSA.

With its tenth anniversary approaching, ICF is grappling with the consequences of its success. "There are a lot of people who would like to be a member of this farm. We've got to figure out how—or if—to expand the membership," says co-farmer Andy Jones, explaining that, since 1996, the farm has had more people interested in joining than it can accommodate.

At the same time, the CSA is considering ways to improve wages and working conditions for its farmer employees, and to remain accessible, in both financial and human terms, to the broader Burlington community. "We would like more revenue. But we need to balance that with what makes the farm feel like a friendly, exciting place to be," Jones explains.

Intervale Community Farm
Burlington, Vermont

Approximately 60 percent of ICF's members are Burlington residents. The rest live within about fifteen miles of the farm in the "bedroom" communities that surround Burlington. Someone from most member households works in Burlington. Approximately 10 percent of the members are lower-income families who receive subsidized shares.

Because of the farm's central location in Burlington, members, as well as the general public, often stop by to visit or ask questions. In fact, home gardeners who have no connection with the farm sometimes call with questions about fertilizing plants or controlling pests. In many ways, the Intervale Community Farm is seen as a Burlington community resource. Acker and Jones say it's a benefit and a challenge. "There are a lot of interruptions. A lot of unexpected visitors asking questions and making demands on the farmers," says Acker. The flip side is a high level of support and a vibrant feeling to the farm.

One of the unique aspects of the Intervale Community Farm is that it is a small farm in a growing neighborhood of small farms in a city. The farm is located in the Intervale, a unique 1,000-acre piece of open and agricultural land in Burlington's North End. It is leased from an umbrella nonprofit organization, the Intervale Foundation, which has for over a decade nurtured a number of other sustainable agriculture enterprises in the Intervale. The ICF, like the other enterprises, leases its land from the foundation, which in turn leases it from both a private owner and the city of Burlington.

The Intervale Foundation's goals are to provide a significant amount of Burlington's fresh food through sustainable farming methods, and to foster agricultural enterprises through which farmers can make a livelihood. Its strategies, explains Will Raap, a founding IF board member, are to acquire long-term control over Intervale land; to build the farmers' capacity to farm, both through infrastructure and education; and to open markets for these farmers.

The base for these efforts is the Intervale compost facility. Launched around the same time as the Intervale CSA, the compost facility is now a joint public-private venture that composts manures and bedding, leaves and food waste. Its profits from tipping fees and some compost sales support other Intervale Foundation initiatives. Moreover, the facility's product, ten thousand tons of compost annually, is used to restore the Intervale's sandy soils. According to Raap,

> The compost project was the logical place to start. We were looking to renew land that was wonderful land, but had been depleted over the last fifty years. Because public officials were looking for cost-effective recycling and waste management alternatives, it also provided a way to get agriculture onto the policy agenda of Burlington.

The Intervale Foundation supports the agricultural enterprises through its "incubator" and "enterprise" farm programs, through which these operations pay subsidized rates for land rental and other technical and in-kind services such as irrigation, access to greenhouse space, use of equipment and machinery, and compost purchases. Additionally, the Intervale Foundation uses some of the compost to prepare land for new growers or for the expansion of existing Intervale enterprises.

In 1998, in addition to ICF, there were four other mixed vegetable farms in the Intervale, as well as cut flower, medicinal herb, and perennial enterprises. Among ICFs neighbors is a new CSA, run by a former ICF apprentice, and to which ICF in 1998 sent its overflow. The Intervale growers have formal monthly meetings, but also try to get together for lunch once a week to share experiences and provide support for each other. "It's really great to be a farmer in a community of farmers," Jones says.

Another neighbor is the foundation's own vegetable operation, Green City Farm (GCF), which is pioneering CSA-like arrangements with several large Burlington-area institutions and extending opportunities to the other Intervale farms to participate in these arrangements. Green City grows vegetables for Fletcher Allen Medical Center, Vermont's largest hospital, and for the New England Culinary Institute, among others. Its relationship with its institutional buyers has evolved over the years.

Motivated by the economics of waste disposal, the hospital, for example, first contracted with the Intervale Foundation to compost its food waste. That relationship paved the way for the hospital's food service to begin buying Intervale-grown produce, first on a purchase-by-purchase basis, and later through contracts.

By 1998, says Raap, the contract between the hospital and GCF provided for GCF to supply a given quantity of food to the hospital, and for the hospital to make monthly payments, in advance. The crop mix is somewhat less diverse than other Intervale vegetable operations, focusing on the crops that the buyers want and that ICF can produce cost-effectively.

With the Intervale Community Farm, the compost project, and the other ventures of the Intervale Foundation, the Green City Farm is a successful example of how the CSA model can help urban communities make food as important as housing, transportation, and employment.

THE COOK COLLEGE STUDENT ORGANIC FARM

CSA as Experiential Learning

*Michael W. Hamm, Ph.D.**

Changes in New Jersey agriculture over the last fifty years are profound and serve as a useful example for agriculture in urbanizing environments. Fifty-one percent of the farmland has been lost to development with agricultural output changing significantly. At the same time, there is a pressing need to develop new farmers capable of growing the food that we depend on for our daily sustenance. New Jersey demographics demonstrate that the average age of a food-producing farmer is approximately fifty-six and rising. This presents a problem in the near future, with fewer farmers to utilize shrinking agricultural land.

In this context, a group of students and faculty founded the Cook College Student Organic Farm in the fall of 1993. The goals of the farm are to give students a place to practice vegetable production within the framework of the state land-grant institution; to provide a context in which students have responsibility for all phases of a farm system, including marketing of the produce; to couple production to

* Michael Hamm is one of the founding faculty sponsors of the Cook College Student Organic Farm at Rutgers in New Jersey. He teaches courses in nutrition and food systems, and is active with NOFA-NJ.

the issue of hunger in New Jersey; to be financially self-sustaining; and to create an opportunity for students to develop group activity and leadership skills.

Starting a student organic farm at an academic institution, even a land grant college, provides several challenges. In our experience, the students, none of whom have had any farming experience, often harbor romantic and unrealistic ideas of what it means to produce crops for market, while being convinced that organic production is the way to grow. Faculty and administrators have years of training, research experience, and/or personal biases that may cause conflict with students' concepts. Local farmers may perceive a land-grant institution marketing produce as unfair competition. The challenge for Cook College was to preserve the vision for an organic farm that fit the wishes of the students and enlisted the participation of faculty, staff, and administrators, while at the same time developing a mechanism for marketing the produce that would not compete directly with local farmers. In the climate of today's college economics, the farm has to be largely self-sustaining while providing summer income for the students.

In our experience, the students, none of whom have had any farming experience, often harbor romantic and unrealistic ideas of what it means to produce crops for market, while being convinced that organic production is the way to grow.

We initially investigated two strategies of produce marketing: a roadside stand and a CSA farm. We decided that a CSA farm was the best mechanism for this student farm. Shares could be sold primarily to faculty and staff within the university and secondarily to families in surrounding communities without competing with local farmers for market share. In addition, the university and surrounding community shareholders would form a base of support for the farm within the university.

Our plan was to expand production slowly while developing programmatic skills and financial independence over a period of several years. We began with $1\frac{1}{4}$ acres of crops in 1994, expanding to 3 acres by 1996. We also started the CSA component modestly, with 24 shares at $150 per share the first year, progressing to 116 shares at $225 per share in 1998. The college administration has provided significant support throughout this project. As the CSA has expanded, revenues have risen to fully cover the daily operation of the farm with regular university support used for capital purchases such as a tractor, rototiller, hand-push precision planter, and irrigation equipment.

The Cook College Student Organic Farm is unusual for a student farm in that it has no full-time manager. Every operation, from seed ordering in winter to cover-crop planting in the fall, is dependent for its completion on the undergraduate students under a collective management system. A student farm director and the recently hired part-time student farm advisor provide continuity, education, and oversight. The student management structure has evolved over the last five years as we have gained a greater appreciation for the nuances of a student farm, which include student turnover, experience, and "student seasons," influenced by course schedules and exams.

The greatest challenge in student management of a college farm CSA continues to be the normal turnover of students as they graduate and leave the area. We aim for stability within an inherently unstable situation: We continually recruit students through various activities while having several students (and their experience) carry over from one summer to the next. This has generally worked in the five seasons of the farm.

We have a slightly truncated season—spring planting time coincides with final exams. Because of this, and the fact that we are a student farm with all the potential vagaries of inexperience, we keep our share prices somewhat below other CSAs in New Jersey, charging $225 per share in 1998. The students work to provide 10 to 15 pounds average per week from the first week of June to mid-October. With approximately twenty-eight crops and seventy-five varieties, the farm provides a diverse harvest for the shareholders. Also, every year seems to generate a new wrinkle. Several students this year are passionate about edible "weeds" and have started a "Weed of the Week" program. Each week they have an article in the weekly newsletter, *The Cover Crop,* about chickweed, wild garlic, lamb's-quarters, or another local plant, and how to prepare it. Two of the students harvest the weeds from around the larger research farm and provide them to the shareholders along with the cultivated crops. It has proved a hit with some of the shareholders.

The shareholders are about 40 percent staff and faculty from the university and about 60 percent local residents, recruited primarily through word-of-mouth by their shareholding university neighbors. In addition, the student farm has a commitment to families and individuals with limited resources in the area. Since its beginning, the farm has provided the majority of fresh vegetables used at Elijah's Promise Soup Kitchen in New Brunswick, which serves meals six days a week and has as many as 150 patrons per day. A local foundation purchases shares for distribution to limited-resource families. During the peak of harvest, the CSA donates excess produce to a local food bank. Overall, about 45 percent of the harvest goes to limited-resource families and the other 55 percent to self-paying shareholders. In this way, Rutgers University, through the efforts of its students, is able to make a contribution to the broader community in which it is located.

The Cook College Student Organic Farm provides a wealth of experiential learning that no classroom instruction could duplicate. In one or two years, students gain a strong appreciation for the efforts needed to produce food, whether organically or conventionally. Many begin with romantic visions and end with realistic perspectives. They learn about their own capacities and limitations. They learn organization, communication, group leadership, and many other skills that will be useful throughout their lives, and also about the relationship between farmers and their customers. Some students discover they really do want to grow food as a profession. Others modify their goals to take an agricultural slant on their career paths.

The CSA provides an opportunity for people in the university and surrounding community to eat fresh, organically grown produce while supporting a group of undergraduate students involved in a special project and community service. The CSA generates a great deal of positive publicity for the university.

Any higher-education institution with some land could develop such a program. It really doesn't matter if it is a quarter acre or several acres. The important points are to have a student-run operation that grows food, to develop a mechanism for selling produce that doesn't compete with local farmers, and to link the project to limited-resource populations in the area. Having faculty or staff who can help provide continuity is very useful. The benefits extend broadly within and outside the institution.

The Cook College Student Organic Farm CSA began as an extracurricular activity. We are now working to integrate some of the activities at the farm with academic courses, and to create a coordinated educational program that will include field trips to other farms in the area and experts meeting with the students at their own farm.

We also hope to help other student farms. Imagine the change in public knowledge and support for local farming if colleges and universities around the country had sustainable student farms providing similar opportunities for interested but inexperienced students. My greatest thrill these last five years has been to watch students grow and develop their skills at the farm. Many of them are pursuing careers in sustainable and organic agriculture across the country or abroad. If we are going to infuse our agricultural system with a sustainable, local context, then providing avenues for the inexperienced to pursue a dream, and ground it in reality is a necessary part of the recipe.

RETURNING RELATIONSHIPS TO FOOD: THE TEIKEI MOVEMENT IN JAPAN

*by Annie Main and Jered Lawson**

We went to Japan on a quest to delve into the history, present status, and thoughts on the future of an agricultural movement that began there twenty-six years ago. The Japanese call it *teikei*, or "face-to-face," and it is similar to our concept of community supported agriculture. The original idea behind the movement was to link urban consumers with organic food and farmers, creating relationships surrounding the food and work that supports them both. In visiting the teikei farms in Japan, some that were part of the initial movement in the early 1970s, we hoped to gain knowledge and insight for the continuing development of food communities back home.

The emergence of the teikei system is a continuation of the complex interaction of Japanese agriculture and Japanese culture. The Japanese population has been supported by an unchanged diet of rice, wheat, vegetables, and fish for more than two thousand years. The volume of production of these staple foods was always a vital factor in supporting their population. Farmers practiced specific systems to maintain the soil fertility and production, and passed soil-enriching knowledge of compost and the usage of plants from generation to generation.

When we arrived at Kaneko-san's farm in the Saitama Prefecture, we were welcomed with the obligatory tea and rice crackers. Mrs. Kaneko proudly showed us the various crops growing: the sesame seeds drying in the greenhouse, the six apprentices harvesting for the day's teikei delivery, the dairy cows, and the biogas accumulator used to extract fuel from the cow dung and other agricultural wastes. Their farm goal, Mrs. Kaneko states, "is to become self-sufficient within our own farm and community."

As a child, Mr. Kaneko remembers his family was almost perfectly self-reliant, with farm-grown rice and vegetables, homemade miso, soy sauce, dairy cows, and other livestock for meat and dairy products. This led him to the idea of providing to others the same items of food that their family was enjoying. Mr. Kaneko started their teikei farmer-consumer partnership in 1971 with these ideas. Some of Mr. Kaneko's members have been eating his farm products for twenty-six years. Their families have grown up on organic food. They have created lifelong relationships with the Kanekos, while receiving the "produce bags."

Later that day, we joined the delivery of produce and eggs with Mrs. Kaneko. We drove through recent suburban developments laid out like those familiar to us in California, stopping to drop the produce bags at various members' homes. At some of the sixteen deliveries we simply dropped the new bag and picked up the old. At a few others we stopped to chat briefly with members, and at one we were invited in for tea and crackers. Driving all over town to drop off such a small quantity of produce seemed inefficient, yet the friendships the Kanekos have created with their members appear invaluable to the community.

Mr. Kaneko, like most of the other farmers we talked with, referred to World War II as the marker for the major changes in agricultural policy and the direction of economic development in Japan. Land was privatized and divided into ever smaller holdings, chemical inputs were promoted to boost yields, and urban industries, such as automobiles and electronics, were promoted to rebuild the Japanese economy. As in most industrializing nations, the consequences were migration to the city, and fewer farmers managing scattered parcels of farmland with increased mechanization and petrochemical inputs.

Consumers, usually housewives, who were concerned about food safety and community problems,

* Annie Main and her husband Jeff have been farming together since 1976 at Good Humus Produce in Capay, California. They started a CSA in 1993. Jered Lawson is the Johnny Appleseed (or Robyn Van En) of CSA in California. After interning at the Homeless Garden, he went on to establish CSA West. Annie and Jered travelled to Japan in the summer of 1997.

Ten Principles of Teikei

1. *Principle of mutual assistance.* The essence of this partnership lies, not in trading itself, but in the friendly relationship between people. Therefore, both producers and consumers should help each other on the basis of mutual understanding. This relation should be established through the reflection of past experiences.
2. *Principle of intended production.* Producers should, through consultation with consumers, intend to produce the maximum amount and maximum variety of produce within the capacity of the farms.
3. *Principle of accepting the produce.* Consumers should accept all the produce that has been grown according to previous consultation between both groups, and their diet should depend as much as possible on this produce.
4. *Principle of mutual concession in the price decision.* In deciding the price of the produce, producers should take full account of savings in labor and cost, due to grading and packaging processes being curtailed, as well as of all their produce being accepted; and consumers should take into full account the benefit of getting fresh, safe, and tasty foods.
5. *Principle of deepening friendly relationships.* The continuous development of this partnership requires the deepening of friendly relationships between producers and consumers. This will be achieved only through maximizing contact between the partners.
6. *Principle of self-distribution.* On this principle, the transportation of produce should be carried out by either the producer's or consumer's groups, up to the latter's depots, without dependence on professional transporters.
7. *Principle of democratic management.* Both groups should avoid over-reliance upon limited number of leaders in their activities, and try to practice democratic management with responsibility shared by all. The particular conditions of the members' families should be taken into consideration on the principle of mutual assistance.
8. *Principle of learning among each group.* Both groups of producers and consumers should attach much importance to studying among themselves, and should try to keep their activities from ending only in the distribution of safe foods.
9. *Principle of maintaining the appropriate group scale.* The full practice of the matters written in the above articles will be difficult if the membership or the territory of these groups becomes too large. That is the reason why both of them should be kept to an appropriate size. The development of this movement in terms of membership should be promoted through increasing the number of groups and the collaboration among them.
10. *Principle of steady development.* In most cases, neither producers nor consumers will be able to enjoy such good conditions as mentioned above from the very beginning. Therefore, it is necessary for both of them to choose promising partners, even if their present situation is unsatisfactory, and to go ahead with the effort to advance in mutual cooperation.

Source: Japan Organic Agriculture Association.

began to demand local organic products and additive-free foods. These women, who were primarily responsible for feeding the family, were not only seeking food safety, but also the *kyodatai* or "community" that was lost in the move to the city. Some of the more politicized and educated urban housewives began to organize consumer cooperatives to solve the problems that were too difficult to solve individually, and found local organic farmers willing to cooperate. One consumer group member recalls, "I first wanted good milk for my children, and then I realized we could find other basics, like safe vegetables, from nearby farmers."

Direct consumer-to-farmer systems emerged. With small land holdings, a tradition of family labor, and a government with a strong infrastructure and subsidy program, the potential for individual growth did not exist. Subsequently, farmers joined efforts with other farmers in order to meet consumer produce demands.

The Japanese philosophy of organic agriculture originated in the Buddhist belief about reincarnation called *shin do fuji* or "uniformity of soil and body." A healthy body is inseparable from a healthy soil. In 1971, a new nonprofit organization emerged, Japan Organic Agriculture Association (JOAA), to promote the re-establishment of organic practices, of organic and biological farming, teikei alliances in Japan, and unification among all the new independent partnerships. They assembled a set of guiding principles, which are listed in the box on page 215.

These principles stressing decentralized, grassroots, local economic development were radical for a country that was rapidly moving toward an export-oriented, industrial-commercial-corporate economy. Yet the number of concerned farmers and consumers who began swimming against the tide of a globalizing food system was growing fast, and the teikei movement spread. Mr. Murayama Masaru of JOAA approximated the number of consumer groups connected with organic producers to be five hundred to one thousand. He also said the groups vary in size from fewer than ten to more than five thousand member families.

Our next stop was a visit with the Kimura family, teikei members in Otsu. We spent several days experiencing their city life and how food from "their" farm affects the Kimura family's daily life. With the addition of the three of us to their family, we were seven people squeezed into a two-bedroom apartment on the third floor of a multistory apartment building. Needless to say, our visit made a strong impact on the family living space, especially with two young children. We found it was important to keep order to the day. One of the first orders of the day was preparing the box of produce that had just arrived the day before from their teikei farm, via a six-hour UPS-like truck-ride from the town of Mochizuki. We were excited to see the box because it reminded us of the CSA boxes back home—a mix of produce that came from people familiar to the Kimuras.

Most of our meals had been the cheap and fast buckwheat noodle soups sold in the shinkansen (light rail train) stations. The meals that this family prepared were different. They had whole grains, clean vegetables, fresh tofu, and a more subtle difference in concern for the health of those eating. While we hadn't met their farmer yet, it was clear they knew where and how the food was grown, and that knowledge came served with each dish—truly satisfying.

Our last farm visit was in Mochizuki, to the four-acre farm owned by Mr. Ooki and his wife. We had reached the farm that delivered to the Kimura family in Otsu. It was an important stop to understand the distance, the connection, and the source of that food we had eaten with the Kimuras. This was one farm where we were given the opportunity to work in the fields all day long, to watch the farmers spending their lives tending the soil, paying detailed attention to plant care and harvesting. It was a pleasure to work, to be a part of their life, to give back to such a generous people. We talked about how different this felt from the congestion around the apartments of the people who ate this food—truly contrasting lives and sacrifices on both sides, yet a common bond through food shared.

After fourteen years of organic farming, Mr. Ooki is still concerned about fine-tuning his fertilization program and the balance of his soil, watching the effects and subtly changing his theory and practices as each new year begins. He grows the vegetables for his members' boxes with a macro-biotic diet in mind, stressing the idea of balancing the characteristics of food. His practice comes from the Chinese philosophy of food as energy, that food is categorized as either hot energy or cold energy, yin or yang, which are balanced to keep the blood clean. Mr. Ooki chooses for the weekly harvest not only what is ripe or in abundance, but also the foods that will help the body stay in balance.

Teikei came from a basic belief that a relationship primarily between the farmer and the consumer was the only true way to ensure the integrity of the production and to build healthy rural and urban communities.

The wisdom that came to us through Mr. Ooki and the other teikei farmers and members is the respect for relationships. A teikei member we talked to about his culture's wisdom agreed with our conclusions. He told us about two Japanese words: *nin-gen* and *hito*. *Nin* means "people" and *gen* means "between." Together, they mean "human being." When the Japanese use this word, it means "something between two people," or a relationship that is spiritual. Without the spiritual, you wouldn't be a human being, just a person—eating or sleeping, but not really "living" your life. The word for this is *hito* or "person." The teikei member felt that we are losing the "something special" between people and instead are putting emphasis on making money. The old Japanese people say, "We cannot take anything to the world of death from this world; we can just bring our soul and heart with us."

After twenty-six years, is the teikei movement growing, stagnating, or declining? Can the face-to-face relationships survive the pressures from an industrializing, impersonal food system? Mr. Kimura, of Nanohanohokai Co-op in Kobe, had the answer most farmers and consumers seem to share: that the teikei future doesn't look too good. Support for regional, national agriculture seems to be waning. In Saitama, the Kanekos said that for every one hundred farmers who die or give up farming, only three younger farmers replace them.

Another long-time consumer and organizer, Mrs. Izumi of the Suzurandai teikei group said, "We are getting older, and not so enthusiastic anymore." Her children have grown up too and moved out of the house, so the produce demands in her household have gone down. When asked why her children didn't continue, she responded, "They went into the city and now are too busy, both [husband and wife] working."

As the teikei movement helped build awareness of the needs for safe food, other marketers saw the potential of this new commodity. If Mrs. Izumi's children are too busy for teikei but still want to get organic produce, they can now find it on the shelves of specialty food stores and large supermarket chains, which often obtain a 50 to 75 percent premium on their organic line. Mrs. Hiraishi of Nanohanohokai handed us a full-color four-page brochure describing home delivery of organic vegetables from Dai-maru, the largest department store in Japan.

With the continuing devaluation of a face-to-face connection with agriculture in society, the generational shift and household makeup in teikei members and farmers, and the growing availability of organic produce in supermarkets and other home-distribution schemes, has come a decline in teikei numbers. No one really knows exactly how much of a decline, but most of the groups we visited reported a drop.

Teikei came from a basic belief that a relationship primarily between the farmer and the consumer was the only true way to ensure the integrity of production and to build healthy rural and urban communities. While such a vision has been the inspiration for hundreds of thousands worldwide, the growing forces of globalization and the aging population of Japan create an uncertain future for teikei. This uncertainty casts a doubt for the survival of Japanese small-farm agriculture. Miho Kimura said, "It is the hardest time in all of Japan's history. Farmers were poor in the past, but now the people cannot make a living from farming the land."

We wondered what knowledge and insight we could bring back home that might help keep the future bright for community-supported small-farm agriculture. We learned that, in an effort to keep face-to-face sustainable agriculture growing and feeding a burgeoning population, we would benefit from finding ways to stimulate the connections to the land and respect for the farming occupation and lifestyle, starting especially with our youth. We would benefit from relationships to the land and to each other. We need to understand that what we feed our bodies is essential food for a healthy life. We need consistent educational outreach to a younger, more diverse membership and farmer base, so as not to create a homogeneous group that ages and fades away. Educational and political campaigns to increase value (both economic and social) for community-scale farming are needed, so that, in turn, farmers will be sufficiently rewarded for their efforts and the younger generation will find interest in the occupation. We must continue with the development and dissemination of educational materials that highlight the principles of community supported agriculture and how they differ from just buying "organic" products at the market.

Principles are important in our individual lives as farmers and consumers. We must be mindfully aware as we try to bridge the gap between urban and rural, and mature with the CSA movement. Celebrating the seasons, having work parties, watching a sunset after planting or harvesting a crop, asking for help, visiting schools, having community dances—these are only a few ideas that can help lead to face-to-face relationships in honor of food and farming.

A Kobe teikei member said, "If we stop eating their organic foods, perhaps they will stop growing organic vegetables. So we must increase the consumers and then work hard. I believe our movement will help protect the lands of the Earth, not only in Japan, but all over. For this purpose, I will continue volunteering." Perhaps Mrs. Kaneko expressed it best when she said, referring to their farm and its members, "Teikei [and CSA] will continue because we understand each other."

CSANYTHING 22

The task for a modern industrial society is to achieve what is now technically realizable, namely a society which is really based on free voluntary participation of people who produce and create, live their lives freely within institutions they control, and with limited hierarchical structures, possibly none at all.

—Noam Chomsky, *Language and Politics*

As a "lumper," someone who is always looking for what people or causes have in common (in contrast to "splitters," who like to keep definitions and organizations very pure), I find it hard to say exactly where the outer boundary of CSA lies. Many kindred community efforts are underway to separate goods and services from the distorting forces of the money economy, and, in the process, to create new participatory institutions. People in traditional rural communities and some groups of low-income city-dwellers have networks of their own, enabling them to survive by exchanging work and sharing resources. "De-centrists" around the country are creating local employment trading systems, local currencies, food co-ops and buying clubs, community investment and development credit programs, revolving loan funds and land trusts, worker-owned co-ops and "new wave" farmer co-ops. Leading the way are futurists like Hazel Henderson, with her critique of what counts as the gross national product, and the practical models and technical support offered by the E. F. Schumacher Society, the Institute for Community Economics, Equity Trust, and others. An alternative system is emerging of "economics as if the Earth really mattered," as Susan Meeker-Lowry calls it. CSAs are definitely part of this encouraging ferment, and some closely related offshoots are adopting the CSA model.

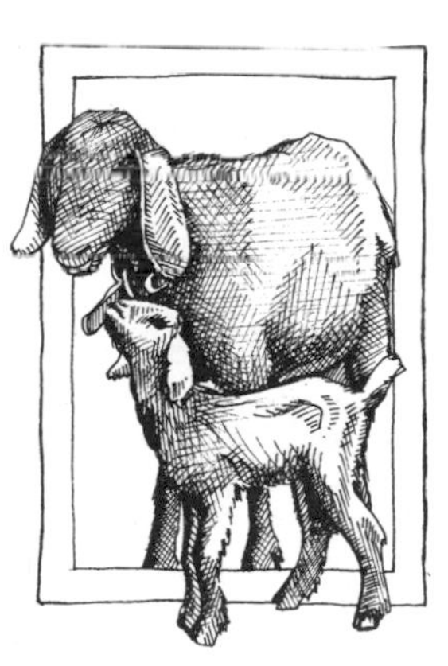

CSSEEDS

Perhaps the most important "input" for food production is the seed from which plants grow. Traditionally, farmers have selected and saved their own seeds, and traded them within farming communities. Producing the variety of seed a CSA farm or garden requires would be a major undertaking. For organic and Biodynamic growers, buying seed that is not treated with chemical fungicides has been an ongoing problem. And, with the introduction of genetically engineered varieties, access to appropriate seed

The Politics of Seed

*by C. R. Lawn**

Do you know where your seed comes from? Probably not, unless you save your own. A handful of giant conglomerates, survivors of the unprecedented industry consolidation of the past thirty years, generate most of the new varieties that show up in the familiar glossy gardening magazines and retail seed catalogs. Even some open-pollinated cultivars such as Blizzard snow peas cannot be propagated for sale, even though the patent holder no longer offers them. Although Sugar Snap peas are offered by hundreds of retailers, all originated from crops grown for Gallatin Valley Seed Co., a subsidiary of Rogers Seed Co., now part of Novartis. All Red Sails lettuce comes from Petoseed, now part of Seminis.

Few seed retailers produce many of their own seed crops. That carrot seed in the Fedco catalog wasn't grown in Maine, nor was seed for that luscious broccoli variety in Vesey's catalog grown on Prince Edward Island. Even independent small seed companies such as Johnny's, Fedco, and Pinetree in my state or Territorial across the continent, are heavily dependent on huge wholesalers. Their retail catalogs are loaded with high-powered Novartis and Seminis varieties such as Silver Queen sweet corn, Jade beans, Sugar Ann snap peas, Celebrity and Big Beef tomatoes, Green Ice lettuce, and Burpee's Golden beet—all industry standards that most commercial growers, whether conventional or organic, would not want to be without. The best retailers run trials in plots located in their climatic region, but grow little of their own seed, instead repacking seed grown by others.

A contrapuntal trend to the consolidation has arisen in the spread of small, quirky regional seed companies. The Seeds Savers Exchange, as well as several of these alternatives, arose to fill the void created by the threatened extinction of many of the open-pollinated varieties. These organizations recognize that unless seed for old open-pollinated and regionally adapted varieties is produced in *commercial* quantities by alternative growers, it will become less and less available in the marketplace. The most idealistic of these new-generation seedpeople, such as Tom Stearns of High Mowing Farm in northern Vermont and Dan Jason of Salt Springs Seed in British Columbia, raise all their own seed in their own locale. Others, serving more commercial growers or mainstream gardeners, are building networks of seed growers to increase the proportion of heirloom and regional varieties in their selections. Alternative breeders such as Tim Peters and Frank Morton in Oregon are beginning to fill a significant niche for limited-volume specialty crops ignored by the wholesalers but revered by an emerging market of organic, Biodynamic, and CSA farmers. Fedco wants half of its new introductions in the next five years to be heirlooms or open-pollinated varieties produced either by staff or other small-scale growers outside the conventional mega-wholesaler network. These emerging trends introduce a ray of hope into an otherwise bleak picture.

At least some of the wholesalers wish the retail industry would mimic its wholesale counterpart, with the small operatives getting big or disappearing. They derisively refer to the entire small farm market as "home gardeners," who should get the bottom of the barrel in quality and service. One such wholesaler, rebuffing Fedco's bid to become a distributor, dismissed the entire market as insignificant, claiming that one of its West Coast grower-processor customers bought as much seed in one year as Fedco could sell to its entire customer base in twenty years! Such a humbling comment puts in stark relief the Lilliputian proportion of the entire alternative agricultural movement when compared with the scope of agribusiness.

* C. R. Lawn, founder of Fedco Seeds, has a good perch from which to inform us of what is happening in seed production and distribution.

Tiny though the movement may be, it is influential and sometimes sends tremors through the big boys. The trend toward heirloom tomatoes has disturbed giants like Petoseed, which has conducted an aggressive advertising campaign trumpeting the virtues of hybrids while denigrating open-pollinated varieties.

Peto, one of the few giants committed to the home garden trade, is well aware of the unpopularity of genetic engineering among home gardeners. Although they are a major player in the biotech game, they took great pains in their 1998 newsletter to assure retailers that their new hybrid Keepsake tomato was "not a transgenic or biotech tomato." Similarly, the All America Selections committee is not currently accepting transgenic varieties as contest entrants, claiming it lacks the resources to educate home gardeners about the benefits of genetically engineered varieties.

Consequently, genetic engineering, while revolutionizing production patterns in cotton, soybeans, canola, and field corn, will make slower inroads in the garden vegetable kingdom. This delay affords the alternative agriculture community, small seed companies, and seed savers a little more time to reduce our dependency on the big boys. Cooperation between small seed companies and heirloom seed savers in the production of seed crops is inevitable and has already begun. Over time, by rebuilding a base of regional specialty varieties, we may be able to secure a measure of independence and carve an enduring, if limited, niche in the marketplace. Probably the best-case scenario in the years ahead is to create parallel food systems in which a significant portion of the population will be able to afford organic produce free from genetically altered organisms.

will become more and more difficult. (For the details of the disturbing trends towards gigantism and corporate control in seed production and distribution, see the essay by C. R. Lawn of Fedco Seeds at left.)

The new "Terminator Technology," which will enable seed companies to produce sterile seed that will grow but not reproduce, may prove to be an even greater threat to seed independence. (The Delta and Pind Land Co. developed this technology with funding from USDA! On May 11, 1998, Monsanto bought Delta, acquiring total control.) A few tiny organic seed companies—Butterbrooke Farm, Bountiful Gardens, Seed Savers Exchange, High Mowing Organic Seed Farm, Seeds of Change, and others—are in the seedling stage. Fedco, a worker-consumer owned co-op, carries some organic seed, no treated seed, and offers graduated discounts to encourage people to cooperate in putting together group orders. M&M/Mars (the candy company) just bought Seeds of Change, the largest of these small companies, but it is too soon to know what that will mean. Except for Fedco, these seed suppliers sell small packets of seed for high prices, beyond the means of small-scale farmers. I calculated one year that buying only organic seed would quadruple my seed bill, already over $600, *if* I could get the quantities I needed.

To take back farmer control over this essential part of growing, Hugh Williams of Threshold Farm in Claverack, New York, has initiated Community Supported Seeds. Inspired by the example of Biodynamic seed production in Bingenheim, Germany, Hugh sent out a call to other growers in 1994 to begin a seed cooperative. Seven growers joined Hugh in an "initiative circle" to develop a plan of action. During that first season, they invited farms to join the project for $100 each; the investment was to cover the expenses of purchasing a germination cabinet to test seed, a record-keeping system, a newsletter and other communications, a room for seed storage, field testing, drying facilities, weighing/counting/packaging equipment, and a catalog. By 1998, twenty-one farms were contributing over one hundred varieties of vegetables, culinary and medicinal herbs, and flowers. In exchange, they receive seed credits. Several hundred growers, from home gardeners to farm-scale operations, pay the $20 subscriber fee. Unlike any other seed catalog, Threshold Seeds identifies which farm produced each variety and describes the farms' production methods, type of soil, elevation above sea level, and annual rainfall. Hugh says that a few people have taken up this project in a wonderful way, though his farm still contributes all the work to keep the co-op going. He would like to venture into research based on insight into the nutrition of selected varieties, but he cannot do any more without additional support. J. J. Haapala, an organic farmer active with Oregon Tilth, has received a grant from the Fund for Rural America to start a similar cooperative venture to grow open-pollinated seed.

NOT BY BREAD ALONE

LOCAL FARM IN CORNWALL BRIDGE, Connecticut, does not sell shares, but, as farmer Debra Tyler says, it is "*community supported* and unquestionably *agriculture*." This NOFA-CT–certified organic dairy is one of eight in the state licensed to sell raw milk, which it bottles in returnable glass. The ten Jersey cows live by rotational grazing. Debra also produces the outrageously pun-filled *Local Moospaper*, "published to keep family and friends posted on the tails and ruminations of Local Farm."

Debra supplied this account of Local Farm's activities to CSA Farm Network:

> Startup capital was raised by selling "milk money" coupons, which were then redeemed for milk after I was in business. Funds for my first shipment of bottles were raised by selling Local Farm tee shirts and a matching donation by a local supporter. Three former customers receive free milk in exchange for delivering to stores near their homes. Coupons are still traded for goods and services. Customers pick up their standing orders of milk at the farm on Saturdays. To discourage traffic to the farm and to encourage neighborly cooperation, we offer a significant price break on orders greater than 3 gallons—this leads to the formation of mini "buying clubs." We offer farm tours for ten bottle-caps and have a "Grand Manure Giveaway" one day each year. We also have potlucks for our Saturday customers.

Local Farm
Cornwall Bridge, Connecticut

SCOTT CHASKEY AND TIM LAIRD of Quail Hill Farm entitled their project to get CSA members to return food scraps to the farm "Community Supported Composting." Suggested to them by Will Brinton of Woods End Laboratory in Temple, Maine, the project received SARE funding under the farmer grant category. Their goals were to create a practical on-farm composting model as a demonstration to other farmers, and as a service to their community. To make member participation easy and odorless, the farm distributed specially designed 1½-gallon biodegradable bags with a leak-proof liner for members to use to store and transport food wastes. (These bags are available from Woods End; see CSA Resources.) Two local restaurants that feature Quail Hill produce on their menus also return their food scraps. As part of the grant, Will Brinton advised Scott and Tim on their composting system, and then participated with them in a field day for the general public. Scott estimates that they gave out fifteen hundred bags a year in 1995 and 1996, which enabled them to return fifteen to eighteen thousand pounds of organic matter to the farm each year. Since the grant, bag distribution continues at a slightly lower rate. Many other CSAs with on-farm pickup invite members to help complete the nutrient cycle through composting. As an added benefit to the church where pickup takes place, GVOCSA members contribute their food wastes to the minister's garden compost pile. While projects like these do not bring in money, they serve the essential functions of returning nutrients to the soil and raising community consciousness.

JEAN GIBLETTE OF HIGH FALLS Gardens in New York has given serious thought to initiating a CSA for Chinese medicinal herbs, but decided that it is premature. Getting away from treating the herbs as a commodity is appealing to Jean, who has been collaborating for several years with university specialists in Nanching and Massachusetts to research appropriate cultivars for North America.The complexity of Chinese herbal medicine poses a tremendous challenge to growers. Practitioners use hundreds of plants never grown in this country, prescribing decoctions of combinations of herbs to treat very mild symptoms. Chinese medicine is a well-documented system going back thousands of years, one that is designed to promote and maintain good health. It never suffered the suppression inflicted on Native American and European herbal systems. A great deal of work lies ahead to identify the Chinese herbs that will grow well on our continent and the native herbs that could become their substitutes. At some time in the future, Jean can imagine organizing a cooperative of growers who would presell their herbs to practitioners.

ANY NUMBER OF PRODUCE-BASED CSAs include bread in their shares or sell separate bread shares. On the Rise sells bread all by itself. At the 1997 Northeast CSA Conference, Karen Kerney told the story of the origins of On the Rise in one baker's delivery route, its eighteen years as a collectively run and owned women's business, and its return full circle to its roots, baking for a community of prepaid subscribers. The project grew out of the food co-op movement, and was financed in part by funds from war tax resisters who set money aside for community projects rather than paying the government. The rest of the money came from the original baker's steady customers, who paid for bread in advance. During its years as a bakery, On the Rise produced high-quality, whole-grain breads. To cover all expenses, they had to sell every loaf they produced, a level of success never achieved. In 1996, the remaining members of the collective decided to close as a business. As Karen put it, they have to trust that, like seeds during a forest fire, they will come through, and when there are more people to bake for, they will rise again. In the meantime, Karen and a few others get together once a month to bake in a facility they share with another baker. Before kneading the bread, Karen calls the members and leaves messages that baking is about to begin. They call her back with orders. Every loaf sells for $2.50: members pick up and wrap the bread themselves.

THE CSANYTHING SESSION at the 1997 Northeast CSA Conference ended with a brainstorm on possible products or services. One participant was starting a feasibility study for fish production. Others mentioned clothing, health care, auto repair, aromatherapy, firewood, Christmas trees, and legal services. The session came to an appropriate close when someone suggested services for the dying and funerals, at which Karen Kerney quipped, "We'll bury you cheap in our compost heap!"

CSAS THAT QUIT 23

The goal ever recedes from us. Salvation lies in the effort, not in the attainment. Full effort is full victory.

—Mahatma Gandhi

We're not sure what percentage of CSAs don't last. We have not been able to track down all the CSAs that started and then quit within five years. This chapter is based on interviews with farmers and sharers from a dozen short-lived CSAs.

One common thread in their stories is the failure to develop a strong core group. Several of the farmers who initiated these CSAs never tried to form a core group. Others did make an effort, but either were not able to find sharers who would take responsibility or failed to recruit replacements when their initial supporters moved on. Finding themselves doing all of the growing, distribution, and organizing, they became discouraged. Former sharers in two of these CSAs told me they were overwhelmed by the quantity of food and felt no control over the selection. More member involvement and better communications could have solved these problems.

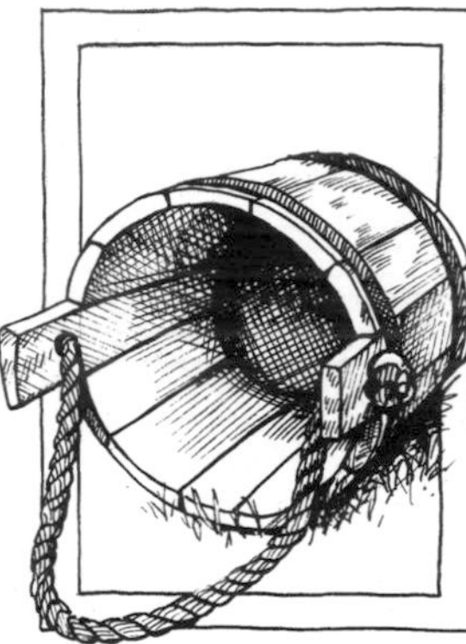

Robert Perry in Homer, New York, blamed himself for failing to "aggressively recruit" new core people after three key members moved away. He said it was a problem "being dependent on labor for weeding till it was overwhelming." Members assumed the CSA would function whether they showed up for their work shifts or not. Robert still believes that his five-year CSA stint was the most successful and satisfying marketing he has done. Jess and Suzanne Unger of Brook Farm in Maryland gave up after two years, blaming too much turnover, problems with recruiting, and the excessive amount of labor required to grow fifty crops. "There was a tremendous amount of running around," said Unger, who made weekly deliveries to Washington, D.C., and Leesburg and Vienna, Virginia. Too late, they realized that a core group might have helped them (*Washington Post*, Sept. 3,1995).

The most frequent reason given for the dissolution of CSAs is not particular to the CSA structure, but rather to the scale of the farm enterprise: the farmer finds a better-paying job. CSAs of only fifteen or twenty or even forty shares cannot offer the farmer enough money to compete with the benefits of a full-time job off the farm. One farmer I talked to decided to concentrate on a different, more promising, form of marketing. In the original Madison Area CSA Coalition (MACSAC) of eight farms, one switched back to other markets after one year of CSA.

Divorce or other emotional upheaval in the farm family destroyed several former CSAs. One such CSA offered the sharers the possibility of using the farm's land to grow food for themselves, but they were unable or unwilling to take on that much responsibility.

Farms in remote rural areas have had difficulty recruiting local sharers. Jim Gerritsen of Wood Prairie Farm in Maine was able to attract newcomers to the county as sharers, but not long-time residents. A few years back in New Mexico, Valerie Kaeppler had a hard time selling the concept of CSA to her rural neighbors. Farmers in other parts of the country similarly report that few of their rural neighbors care to purchase organically grown foods. In these areas, chemically sensitive people or hard-core organic advocates who lack the time to garden are the most likely candidates. Overall, it does seem easier to find potential CSA members in cities or suburbs. Adapting to the needs of local people is critical, however: there are examples of thriving projects in rural Vermont and Iowa, where farmers have been able to find the right mixture of CSA and other markets, or have targeted groups such as the elderly and provided home delivery.

The most frequent reason given for the dissolution of CSAs is not particular to the CSA structure, but rather to the scale of the farm enterprise: the farmer finds a better-paying job.

Two of the CSAs I interviewed fell apart because the grower lacked farming experience, resulting in inadequate production, disorganization, and poor quality. One grower tried to get started in a year of a serious drought, undercharged for his shares, and then could not produce enough food even for that limited payment. He gave up in August. Neither he nor the two growers for the other CSA had any previous experience in growing for market. Before allowing a new CSA to join their network, the MACSAC growers ask new farmers to fill out a questionnaire on their practices and previous experience. If they lack experience and knowledge, MACSAC encourages them to intern or apprentice with another farm. If their production is too specialized, MACSAC advises developing cooperative relationships with existing CSAs (Margie Ostrom, "Community Farm Coalitions," in *Farms of Tomorrow Revisited*, p. 94).

Jim Volkhausen of Buttermilk Farm near Ithaca, New York, stopped doing a CSA to take another job. He had other interests, and with only forty shares could never have lived on just the farm income. Other CSAs in the area picked up his members. Despite its failure to produce enough money, Jim considered the CSA successful: "The concept of member involvement was very important, and I would encourage others to do it that way for the sense of community."

After six years of successful operation, Pat and Mike Kane of Shamrock Hill Farm in Port Crane, New York, decided to take a year's sabbatical from the CSA. Pat's non-farm job had become too consuming to allow her much time to help Mike on the farm, and Mike couldn't do the high-quality job he likes by himself. Their sharers were very supportive, and agreed to keep the core together through the year off by focusing on bulk buying. Farmers and sharers hoped to redesign the project for the future to make it easier for the farm. However, after another year, when Pat's job continued to expand, they regretfully gave up on the CSA.

The failure to create a real sense of community participation led Harvey Harman at Sustenance Farm near Raleigh-Durham, North Carolina, to disband his CSA. For five years, the project worked well enough; members were happy to get the vegetables and to support the farm financially, though the forty shares generated too little income to provide a living for Harman. With thirty to forty miles between their homes and the farm, the members never became active. Turnover from year to year was high. The core group was forced to spend too much of its energy recruiting new members. Even social events, such as potlucks, which the farm sponsored in town, were not well attended. As Harvey put it, the CSA was a good thing, but "the community aspect never happened." Two years later, Harvey is planning to give CSA another try.

Marty Rice, of Country Pleasures Farm in Maryland, told a reporter for the *Washington Post* that she gave up on CSA because she had to work 80 to 90 hours a week to gross $6,300 for the year. "We did an end-of-season survey, and asked people if they would be willing to have smaller shares, pay more, or contribute some labor, and the answer was no, no, no. People didn't have an understanding of the concept, that we're in this together," Marty complained, but concluded, "It's a very important idea, but we have to figure out how to make it succeed" (*Washington Post*, Sept. 3, 1995).

SOME ATTRITION OF CSAS IS INEVITABLE. The same pressures that wear down other farms (and many other small, family-run businesses) do their destructive work on CSAs as well. And, of course, CSA does not suit every farmer or every consumer. In too many cases, though, farmers have given up on CSAs because they did not know how to organize the support from the members that they needed, or members either did not understand or know how to give the help that would have kept their farmers going. Hopefully this book, with its outpouring of generous sharing of experiences and bright ideas, will save a few shaky CSAs from going over the edge.

THE FUTURE: ON ACTIVE HOPE

When asked what he would do if the world were to end tomorrow, Martin Luther reportedly answered, "I would plant an apple tree today."

—"CARING FOR CREATION: VISION, HOPE AND JUSTICE,"
A SOCIAL STATEMENT FROM THE
EVANGELICAL LUTHERAN CHURCH IN AMERICA

Looking around us at the shopping strips and endless parking lots, the undrinkable water, and massive fish kills, listening to the news of children murdering children, of self-serving politicians, of batterings, of senseless car wrecks, and billion dollar buyouts with downsizings in the tens of thousands—it is easy to give in to cynicism or despair. Overshadowed by the enormous forces of the global economy, each of us feels about as powerless as a tiny ground beetle. Reviewing the history of this century, however, Howard Zinn observes that, above all else, its course has been marked by unpredictability. Contrary to all odds, ordinary people have won the most unpromising and unequal of battles. And upon this unpredictability, Zinn offers hope:

> . . . the struggle for justice should never be abandoned because of the apparent overwhelming power of those who have the guns and the money and who seem invincible in their determination to hold on to it. The apparent power has, again and again, proved vulnerable to human qualities less measurable than bombs and dollars: moral fervor, determination, unity, organization, sacrifice, wit, ingenuity, courage, patience—whether by blacks in Alabama and South Africa, peasants in El Salvador, Nicaragua and Vietnam, or workers and intellectuals in Poland, Hungary, and the Soviet Union itself. [And I would add, by people all over the world to prevent a nuclear holocaust.] No cold calculation of the balance of power need deter people who are persuaded that their cause is just. (*The Zinn Reader*, p. 642)

We can go on eating our way to obesity, heart attacks, cancer, infertility, and total dependence on the Philip Morrises who dominate our food supply, or we can start to take charge of what we put in our mouths. Our tiny groups of farmers and aware eaters have the qualities Zinn lists, and we have a cause as just and vital as any for which people have ever struggled. Buying locally grown food is a first step towards the health of our own bodies, and the health of our local communities. Joining in community supported agriculture is another step toward nurturing the interdependence among humans, the soil, the plants, and other creatures that we must have if we hope for a future on this very small planet.

As Robyn Van En put it in her notes for this conclusion, "CSA is a viable contender to the reckless and unsustainable food system to which we have grown accustomed. CSA strives to be socially and ecologically responsible, to educate and empower, while providing good food, one of the basic necessities of life. . . . It is a participatory means to securing your food supply for today and for future generations."

CSA farms and gardens around the country are changing and growing. The Genesee Valley Organic CSA is moving to new ground, while exploring legal structures to give members more control. Most members' work for 1998 will be on design, barn renovation, and greenhouse construction to be ready for crops in 1999. Like Fairview Gardens, Silver Creek and Angelic Organics are initiating educational programs to give more inner-city youngsters the chance to learn about rural life and the ecological growing of food. Be Wise Ranch is modernizing its packing facility and designing a retail stand to move more of its production through direct sales. Holcomb Farm is shifting from dependence on grant money to financial self-sufficiency. Common Harvest and Roxbury Farms are expanding their roles as distribution centers for local products from other farms, while value-added processing both by individual farms and co-ops is spreading in New York, Kansas, and Appalachia. Genesis Farm plans to grow

MARILYN ANDERSON

Elizabeth Henderson bunching turnip greens in the rain at Rose Valley Farm.

grains and begin baking its members' daily bread. In rural Iowa, coastal California, and even in the concrete depths of New York City, regional support groups are reaching out to conventional farmers with the promise of customers who will share the risk as the farms reduce their reliance on agrichemicals. In New York and Oregon, farmers are increasing the production and sharing of seed. Lower-income people are no longer an afterthought, but active participants in more and more CSAs in California, New Jersey, and Montana. And at the Denver airport, in central Philadelphia, in Tacoma, and in Brooklyn, CSA food production on urban garden sites is underway.

Whether CSAs number ten thousand in the year 2000 is less important than whether the six hundred to one thousand CSAs that already exist mature and flower. Invaluable as computer models and sophisticated indicators of sustainability may be, the real story is happening on the ground in community supported farms and gardens, a living reality for thousands of people who are learning to work together to live more sustainably. In the shadow of the global supermarket, CSAs, like one thousand farmer-controlled experiment stations, are busy with research on the social and economic relations of the future. Each of us may be as insignificant as a ground beetle, but together the ground beetles of the globe keep it from being buried in a thick layer of rot.

In another of his wonderful vegetable love notes to Rose Valley Farm, Josh Tenenbaum wrote:

> Thich Nhat Hahn, the Buddhist monk and peace activist, recommends that we have a great store of positive images to counterbalance the negative ones. And it is easy for me to think of the beautiful food life that you nurture, poetry in fiber and carbohydrate. . . . I know that you do the most fundamental kind of peace work, each and every day on your farm. For this I am speechless with gratitude (well, except for this letter, I am speechless) and deeply honored to know you. You live now in my images, but also in me, from your living energy that goes into your food, and from these into me. I try to keep the chain going, and pass it along to others.

I don't think Josh will mind sharing his image and allowing me to extend it to all the CSAs on this continent and around the globe, as I invite them to share from this book to help build a peaceful world of productive community farms and gardens.

CSA RESOURCES

THE TECHNICAL RESOURCES AVAILABLE to CSAs are increasing rapidly. Organizations around the country are taking on the task of maintaining updated lists of existing CSAs, and distributing or circulating printed materials, slides, and videos. Rather than try to list all of the materials and risk being out of date before the ink dries, this section will list the resource centers to which people interested in CSA can turn for the latest information. Please contact sources directly for current lists and prices of materials.

NATIONAL ORGANIZATIONS

Alternative Farming Systems Information Center
National Agriculture Library Research Service
U.S. Department of Agriculture
10301 Baltimore Ave., Rm. 304
Beltsville, MD 20705-2351
301-504-6559; FAX: 301-504-6409
e-mail: afsic@nal.usda.gov

Bibliographies on CSA, organic farming, and related topics.

Appropriate Technology Transfer for Rural Areas (ATTRA)
PO Box 3657
Fayetteville, AR 72702
800-346-9140
e-mail: askattra@ncatfyv.uark.edu

Packet of basic information on CSA. Also provides packets on such topics as apprenticeships, organic farming, marketing. Staff will either send you a pre-assembled packet or design a packet of information to answer your specific questions. Also have tapes and videos. Great service!

The Biodynamic Farming and Gardening Association
PO Box 550
Kimberton, PA19442
800-516-7797 or 610-935-7797;
FAX: 610-983-3196
www.prairienet.org/psca
Internet chat room: csa-L@prairienet.org (to subscribe, send e-mail message to listproc@prairenet.org stating "subscribe csa-L [your e-mail address]" in the message)

The Biodynamic Association maintains a database of CSAs throughout the country. For a CSA close to you, call for a listing for your state. Their list is only as good as the information we supply them, so if you are starting a CSA or stopping, please let them know.

PUBLICATIONS

It's Not Just About Vegetables, 25 minute VHS video, introducing CSA with Robyn Van En, 1988.

CSA 1, "Community Supported Agriculture," audio tape by grower panel: Decater, Moore, Geiger, Altemueller.

CSA 2, "The Future of Community Support Agriculture," audio tape by Hartmut von Jeetze.

CSA 3, "Farms of the Future—Food of the Future," audio tape by Fred Kirschenmann.

Farms of Tomorrow Revisited, by Trauger Groh and Steven McFadden, 1997, also distributed by Chelsea Green. 294 pp.

Basic Formula to Create CSA, by Robyn Van En. 32 pp. Predecessor to this book.

Louise's Leaves, by Louise Frazier, 1994, a guide to year-round nutrition with seasonal foods and home storage.

"Introduction to Community Supported Farms/Gardens, Farm/Garden Supported Communities," 10 pp. free booklet.

CSA Works
Michael Docter, Linda Hildebrand, Dan Kaplan
115 Bay Rd.
Hadley, MA 01035
413-586-5133

Provides technical assistance to CSA farmers only in the winter (Dec.–Mar.). Reasonably priced consultations on assessing your readiness to start a CSA; help in converting to CSA; source of equipment and materials, such as value-added recipe packets, scrip, crop planning chart. Video on efficient harvesting.

E. F. Schumacher Society
Box 76, RD 3
Great Barrington, MA 01230
413-528-1737
e-mail: efssociety@aol.com
www.members.aol.com/efssociety

Robert Swan and Susan Witt provide advice and information on all the ways small can be beautiful: land trusts, local currencies, sample leases, excellent decentrist library. Many useful booklets on land trusts and local economics, such as "A New Lease on Farmland," "Land: Challenge and Opportunity," "Local Currencies: Catalysts for Sustainable Local Economies."

Equity Trust
539 Beach Pond Road
Voluntown, CT 06384
Phone/FAX: 860-376-6174

Chuck Matthei—with years of experience in land-tenure issues and land trusts—can provide information, a range of flexible solutions, and contacts with sources of funding or local land trusts. Gives stimulating workshops on "Gaining Ground."

Fund for CSA: land tenure counseling, financing through revolving loan fund for CSAs, excellent source of information on land-related issues.

Northeast Regional Agricultural Engineering Service (NRAES)
152 Riley-Robb Hall
Cooperative Extension
Ithaca, NY 14853-5701

Cooperative Extension publishers. Publish excellent technical manuals on irrigation, produce handling for direct marketing, on-farm composting, refrigeration and controlled atmosphere for horticulture crops, greenhouse engineering, and more.

Rocky Mountain Institute
1739 Snowmass Creek Rd.
Snowmass, CO 81654-9199
303-927-3851

Provides research, training, and resource materials on sustainable energy, water, agriculture, transportation, and community economic development.

Sustainable Agriculture Network (SAN)
Andy Clark, Coordinator
National Agriculture Library, Rm. 304
301-504-6425
e-mail: san@nal.usda.gov
www.ces.ncsu/san/

National outreach of the USDA Sustainable Agriculture Research and Education (SARE) Program. Provides publications on sustainable agriculture, information on research projects funded by SARE, videos, chat room, Web site.

Woods End Research Laboratory
Box 297
Mount Vernon, ME 04352
207-293-2457; FAX: 207-293-2488
e-mail: info@woodsend.org

Compost testing and consulting; source of Community Supported Composting bags.

REGIONAL ORGANIZATIONS

The Center for Sustainable Living
Steve Moore, Director
Wilson College
1015 Philadelphia Ave.
Chambersburg, PA 17201
717-264-4141, ext. 3247; FAX: 717-264-1578
e-mail: wc-csl@mail.cvn.net

The Wilson College CSA Project can supply a CSA brochure, which serves as a general introduction to the concept; a CSA information packet; an 80-page "CSA Handbook: A Practical Guide to Starting and Operating a Successful CSA"; and a collection of clip art.

PUBLICATIONS

"CSA: Making a Difference," 15 minute video. Interviews and footage from four CSA farms. Well-made video, provides good introduction to concept.

"CSA: Be Part of the Solution," Slide show in two versions: text slides only—you can add your own farm's slides as illustrations—or text and photo slides.

CSA Farm Network Publications
Steve Gilman
130 Ruckytucks Rd.
Stillwater, NY 12170
518-583-4613
e-mail: sgilman@netheaven.com

Also available from NOFA-NY.

CSA West
Weyland Southon, coordinator
PO Box 363
Davis, CA 95617
530-756-8518; FAX: 530-756-7857
e-mail: csawest@caff.org

Publishes directory of CSAs in California, sets up workshops, provides resource materials and technical advice.

Great Lakes Area CSA Coalition (GLACSAC)
c/o Peter Seely
W7065 Silver Spring Lane
Plymouth, WI 53073
414-8922-4856

Member organization providing networking for eastern Wisconsin and northeastern Illinois CSAs.

Iowa Network for Community Agriculture (INCA)
1465 120th St.
Kanawha, IA 50447
Jan Libbey, coordinator
515-495-6367
e-mail: libland@Kalnet.com

Iowa State University Extension Publications
119 Printing and Publications
Iowa State University
Ames, IA 50011-3171
515-294-5247; FAX: 515-294-2945

Statewide list of CSA farms and organizers.

PUBLICATIONS

Iowa CSA Resource Guide for Producers and Organizers—excellent resource booklet with application far beyond the borders of Iowa, with sections on legal issues, seed, supplies, livestock, flowers, and other products.

Just Food
625 Broadway, Suite 9C
New York, NY 10012
212-674-8124; FAX: 212-505-8613
e-mail: justfood@igc.org

CSA in New York; support project for CSAs in New York City area; helps connect farmers and potential members; guides formation of core groups for CSAs. Good materials on core and farm CSA budgeting and responsibilities.

Land Stewardship Project
2200 4th St.
White Bear Lake, MN 55110
651-653-0618; FAX 612-653-0583

Newsletter, workshops, videos, on-farm training.

Madison Area Community Supported Agricultural Coalition (MACSAC)
c/o Wisconsin Rural Development Center
216 Main St.
Mt. Horeb, WI 53572
608-437-5971; FAX: 608-437-5972

Publishes annual CSA directory, provides support and networking for CSA farms.

Michael Fields Agricultural Institute
W2493 County Road ES
East Troy, WI 53102
414-642-3303; FAX: 414-642-4028

Publishes and distributes "Upper Midwest Organic Resource Directory," and "CSA: Upper Midwest Regional Directory"; includes area resources and CSA networks. Distributes "From Asparagus to Zucchini: A Guide to Farm-Fresh, Seasonal Produce."

NOFA-NY
PO Box 21
S. Butler, NY 13154
315-365-2299; FAX: 315-365-3299
e-mail: nofany@ny.tds.net

Northeast Sustainable Agriculture Working Group (NESAWG)
PO Box 608
Belchertown, MA 01007-0608
413-323-4531; FAX: 413-323-9595
e-mail: nesfi@igc.apc.org

Currently serving as the CSA resource center for the Northeast, replacing CSA North America. Can provide lists of CSAs, contacts, sources of additional resource materials.

Rural Development Center (RDC)
PO Box 5415
Salinas, CA 93915-5415
408-758-1469; FAX: 408-758-3665
e-mail: rdcfarm@aol.com

Southern Sustainable Agriculture Working Group
PO Box 324
Elkins, AR 72727-0324
501-587-0888
e-mail: ssfarm@juno.com

Keith Richards, staff person, can help locate CSAs in southern states.

PUBLICATIONS

Making It on the Farm: Increasing Sustainability through Value-Added Processing and Marketing. 40 pp.

PERIODICALS AND PUBLICATIONS

Biodynamics: A Bimonthly Magazine Centered on Health and Wholeness
Biodynamic Association
PO Box 550
Kimberton, PA 19442
800-516-7797

6 issues/year, $35.

The Community Farm: A Voice for CSA
3480 Potter Rd.
Bear Lake, MI 49614
616-889-3216
e-mail: fsfarm@mufn.org

Quarterly newsletter, $20/year subscription ($23 in Canada, $28 overseas).

Dollars and Sense: What's Left in Economics
One Summer St.
Somerville, MA 02143
617-628-8411

Bimonthly, $22.95/year. Analyses of current economic goings-on in the U.S. and around the world in language normal people can understand.

A FoodBook for a Sustainable Harvest
by Elizabeth Henderson and David Stern, 1994

To order copies, write to Elizabeth Henderson at Peacework Organic Farm, 2218 Welcher Road, Newark, NY 14513; $11/copy plus $2.75 postage and handling.

GEO (Grassroot Economic Organizing Newsletter)
PO Box 5065
New Haven, CT 06525
203-389-6194

Bimonthly, $15/year. Covers worker-owned enterprises, co-ops, community-based businesses, community-labor environmental coalitions.

Growing for Market: News and Ideas for Market Gardeners
Fairplain Publications
PO Box 3747
Lawrence, KS 66046
785-748-0605, 1-800-307-8949

Monthly, $27 subscription. Carries regular articles on CSAs and related small farm topics. Also sells "Marketing Your Produce: Ideas for Small-Scale Farmers," collection of best marketing articles, 1992–1995.

The Natural Farmer
NOFA
411 Sheldon Rd.
Barre, MA 01005
978-355-2853

Quarterly, free with NOFA membership; otherwise $10/year. NOFA also sells video tapes of conference workshops on a wide range of topics related to organic farming, gardening, marketing, lifestyle, and distributes "The Real Dirt: Farmers Tell About Organic and Low-Input Practices in the Northeast," and the five Northeast Farmer to Farmer Information Exchange booklets on organic apples, sweet corn, strawberries, greenhouse, and livestock.

To Till It and Keep It: New Models for Congregational Involvement with the Land
by Dan Guenthner

Available from Job Ebenezer, Division for Church in Society, Evangelical Lutheran Church in America, 8765 Higgins Road West, Chicago, IL 69631; 800-683-3522 ext. 2708.

Organizing Manual for Social Change: A Manual for Activists in the 1990s
Arlington, Virginia: Seven Locks Press, 1991

Lists organizations and how to contact them. Available from the publisher at 800-354-5348.

OTHER ORGANIZATIONS

Center for Holistic Management
1010 Tijeras NW
Albuquerque, NM 87102
505-842-5252; FAX: 505-842-5252
e-mail: center@holisticmanagement.org

Provides training and resource materials on Holistic Management. Distributes "Holistic Resource Management: A New Framework for Decision Making"; $30 paperback, $50 hardcover.

Center for Rural Affairs (CRA)
PO Box 406
Walthill, NE 68067
402-846-5428; FAX: 402-846-5420

The CRA newsletter is best source in this country for understanding what is going on in agriculture. The Center has projects in rural economic development, sustainable agriculture, and agricultural policy on local, state, and national levels.

Farming Alternative Program
Department of Rural Sociology
Warren Hall
Cornell University
Ithaca, NY 14853
607-255-9832
e-mail: jmp32@cornell.edu

Resource materials and publications on community agriculture development, farmers' markets, agritourism, feasibility of new farm enterprises, small-scale processing.

International Forum on Globalization
1555 Pacific Ave.
San Francisco, CA 94109

Newsletter, books, reports, tapes on corporate rule and how to change it.

Seed Savers Exchange
3076 North Winn Road
Decorah, IA 52101
319-382-5872

Annual membership fee of $25 includes Yearbook through which participating members can offer nonhybrid vegetable seeds and order from other members. Other publications on seed saving; also sell selected heirloom seeds in small quantities.

A Whole New Approach
Ed Martsolf
1039 Winrock Drive
Morrilona, AR 72110
Phone/FAX: 501-727-5659
e-mail: ed.martsolf@mev.net

Provides training in Whole Farm Planning.

SUPPORT FOR FARMER RESEARCH

Northeast Region Sustainable Agriculture Research and Education Program (SARE)
10 Hills Building, Carrigan Drive
University of Vermont
Burlington, VT 05405-0082
802-656-0471; FAX: 802-656-4656
www.uvm.edu/nnesare/

Northeast SARE offers two kinds of grants to producer-initiated and -managed projects: Farmer/Grower Grants to conduct farm-based experiments to answer production and marketing questions; and SEED (Special Evaluation, Education and Demonstration) grants to farm-test selected, alternative *practices. Grants will be awarded on a competitive basis to farmers in the twelve-state region (Connecticut, Delaware, Maine, Maryland, Massachusetts, New Hampshire, New Jersey, New York, Pennsylvania, Rhode Island, Vermont, Washington, D.C.). In the past, grants have ranged from $300 to about $8,000. Grant applications were due December 11 in 1998; call, write, or visit the Northeast Region SARE Web site for information.*

Organic Farming Research Foundation
PO Box 440
Santa Cruz, CA 95061-9984
408-426-6606; FAX: 408-426-6670
e-mail: research@ofrf.org

OFRF has two funding cycles per year. Grant application deadlines are July 15 and January 15. Projects may be farmer-initiated, or should involve farmers in both design and execution and take place on working organic farms whenever possible. Modest proposals of $3,000 to $5,000 are encouraged. Contact OFRF office for procedures and grant applications. A complete list of OFRF-funded projects is also available upon request.

REFERENCES

Ableman, Michael. *On Good Land: The Autobiography of an Urban Farm.* San Francisco: Chronicle Books, 1998.

Alaimo, Katherine. "Food Insufficiency Exists in the United States: Results from the Third National Health and Nutrition Examination Survey (NHANES III)." *American Journal of Public Health* 88, no. 3 (March 1998): 419–26.

American Farmland: The Magazine of American Farmland Trust (Summer 1998): 4 (Letters: editor's note).

Ashman, Linda, et al., "Seeds of Change: Strategies for Food Security in the Inner City." Master's thesis, University of California, Los Angeles,1993. Available from the Community Food Security Coalition, PO Box 209, Venice, CA 90294; 310-822-5410.

Berry, Wendell. "Conserving Communities." In *The Case Against the Global Economy and for a Turn Toward the Local,* ed. Jerry Mander and Edward Goldsmith. San Francisco: Sierra Books, 1996.

Brown, Lester R. *State of the World 1998.* New York: W.W. Norton, Worldwatch Institute, 1998.

"Cash Receipts and Farm Income." *New York Agricultural Statistics* (September 1995).

Cohn, Gerry. "Community Supported Agriculture: Survey and Analysis of Consumer Motivations." Research paper, University of California at Davis, 1996. Available from Gerry Cohn, 127-B Hillside St., Asheville, NC 28801.

Colborn, Theo, Dianne Dumanoski, and John Peterson Meyers. *Our Stolen Future: Are We Threatening our Fertility, Intelligence, and Survival?* New York: Dutton, 1996.

Daly, Herman E. "Sustainable Growth? No Thank You." In *The Case Against the Global Economy,* ed. Jerry Mander and Edward Goldsmith. San Francisco: Sierra Books, 1996.

Diehl, Janet. *The Conservation Easement Handbook.* Boston: Land Trust Exchange, 1988.

Duesing, Bill. "Holy Day Connection." Broadcast on "Living on Earth," National Public Radio, April 10, 1998.

Dunaway, Vicki. "The CSA Connection." *FARM: Food Alternatives with Relationship Marketing* (Summer/Fall 1996). 1–6.

Feder, Barnaby J. "Getting Biotechnology Set to Hatch," New York Times, 2 May 1998, pp. D1, D15.

Freegood, Julia. Presentation to Farmland Preservation workshop, SARE Tenth Anniversary, Austin, Texas, 5–6 March 1998.

Gilman, Steve. "Our Stories," *CSA Farm Network,* vol. II, pp. 81–83. Stillwater, N.Y.: CSA Farm Network, 1997.

Goldberg, Ray. "New International Linkages Shaping the U.S. Food System." In *Food and Agricultural Markets: The Quiet Revolution*, ed. Lyle P. Schertz and Lynn M. Daft. NPA Report #270. Washington, D.C.: Economic Research Service, USDA, Food and Agriculture Committee and National Planning Association, 1994.

"Green Greens? The Truth About Organic Food," *Consumer Reports* (January 1998): 12–17.

Grubinger, Vern. *Sustainable Vegetable Production on One Hundred Acres.* Ithaca, N.Y.: NRAES, 1998.

Guebert, Alan. "NAFTA Is Proving to be a Disaster." *Agri News*, 9 October 1997.

Heffernan, William D. "Domination of World Agriculture by Transnational Corporations (TNCs)." In *For ALL Generations: Making World Agriculture More Sustainable*, ed. J. Patrick Madden and Scott G. Chaplowe, pp. 173–81. Glendale, Cal.: WSAA, 1997.

Hightower, Jim. Keynote speech, SARE Tenth Anniversary, Austin, Texas, March 5, 1998.

Hightower, Jim. *There's Nothing in the Middle of the Road But Yellow Stripes and Dead Armadillos.* New York: Harper Collins, 1997.

Hoffman, Judith. "CSA: A Two-Part Discussion." *Small Farmers Journal* 19, no. 4 (Fall): 28–29.

Institute for Community Economics, *The Community Land Trust Handbook*. Emmaus, Penn.: Rodale, 1982.

Osterholm, Michael, Minnesota State Epidemiologist. Interview on "Fresh Air," National Public Radio, 5 May 1998.

Kane, Deborah J., and Luanne Lohr. "Maximizing Shareholder Retention in Southeastern CSAs: A Step Toward Long-term Stability." University of Georgia, 1997.

Kelvin, Rochelle. *Community Supported Agriculture on the Urban Fringe: Case Study and Survey.* Kutztown, Penn.: Rodale Institute Research Center, 1994.

Kittredge, Jack. "Community Supported Agriculture: Rediscovering Community." In *Rooted in the Land: Essays on Community and Place*, ed. William Vitek and Wes Jackson. New Haven: Yale University Press, 1996.

Kittredge, Jack. "CSAs in the Northeast: The Farmers Speak." *The Natural Farmer* (Summer 1993): 12.

Kolodinsky, Jane, and Leslie Pelch. "Factors Influencing the Decision to Join a Community Supported Agriculture (CSA) Farm." *Journal of Sustainable Agriculture* 10, no. 2/3 (1997): 129–41.

Kolodinsky, Jane. "An Economic Analysis of Community Supported Agriculture Consumers." SARE grant #5-24544, 1996.

Korten, David. *When Corporations Rule the World.* West Hartford, Conn.: Kumarian Press, 1995.

Laird, Timothy J. "Community Supported Agriculture: A Study of an Emerging Agricultural Alternative." Master's thesis, University of Vermont, 1995.

Lass, Dan, and Jack Colley. "What's Your Share Worth?" *CSA Farm Network,* vol. II, pp. 16–19. Stillwater, N.Y.: CSA Farm Network, 1997.

Lehman, Karen, and Al Krebs. "Control of the World's Food Supply." In *The Case Against the Global Economy,* ed. Jerry Mander and Edward Goldsmith. San Francisco: Sierra Books, 1996.

Lewis, W. J. et. al. "A Total System Approach to Sustainable Pest Management." *Proceedings of the National Academy of Sciences* 94 (11 November 1997): 12243–48.

Looker, Dan. *Farmers for the Future.* Ames: Iowa State University Press, 1996.

Lyson, Tom. "The House that Tobacco Built." *Food, Farm and Consumer Forum* 6 (January 1988): 2.

McFadden, Steven, and Trauger Groh. *Farms of Tomorrow Revisited: Community Supported Farms—Farm Supported Communities.* Kimberton, Penn.: BD Association, 1997.

MacGillis, Therese. "Food as Sacrament." In *Earth and Spirit: The Spiritual Dimension of the Environmental Crisis*, ed. Fritz Hull. New York: Continuum, 1993.

Markley, Kristen. "Sustainable Agriculture and Hunger." Master's thesis, Pennsylvania State University, 1997.

O'Brien, Patrick. "Implications for Policy." In *Food and Agricultural Markets: The Quiet Revolution*, ed. Lyle P. Schertz and Lynn M. Daft. NPA Report #270. Washington, D.C.: Economic Research Service, USDA, Food and Agriculture Committee and National Planning Association, 1994.

Pennsylvania State Extension, "Guidelines for Renting Farm Real Estate in the Northeastern United States."

Pesticide Action Network of North America. Sustainable Agriculture Network list-serve, PANNA Updates Service, panna@panna.org, May 1988.

"The Rich." *New York Times Magazine*, 19 November 1995, p. 1.

Ritchie, Mark. "The Loss of Our Family Farms." Minneapolis: League of Rural Voters, 1979. Reprinted by Center for Rural Studies, San Francisco, 1979.

Scher, Les, and Carol Scher. *Finding and Buying Your Dream Home in the Country*. Chicago: Dearborn Financial Publishing, 1996.

E. F. Schumacher Society, "A New Lease on Farmland." Great Barrington, Mass.: E. F. Schumacher Society, 1990.

Smith, Stewart N. "Farming Activities and Family Farms: Getting the Concepts Right." Presentation at the Joint Economic Committee Symposium, Washington, D.C., 21 October 1992.

Strange, Marty. *Family Farming: A New Economic Vision*. Lincoln: University of Nebraska Press, 1988.

Strange, Marty. "Peace with the Land, Justice among Ourselves." *CRA Newsletter* (March 1997).

Suppan, Steve, and Karen Lehman. "Food Security and Agricultural Trade under NAFTA." Minneapolis: IATP, July 1997. Available from the Institute for Agriculture and Trade Policy, 2105 First Ave. S., Minneapolis, MN 55404; 612-870-0453.

Suput, Dorothy. "Community Supported Agriculture in Massachusetts: Status, Benefits, and Barriers." Master's thesis, Tufts University, 1992.

USDA. *A Time to Choose: Summary Report on the Structure of Agriculture*. Washington, D.C.: USDA, 1981.

USDA. Economic Research Statistical Bulletin #849, December 1992.

USDA. "Your Farm Lease Checklist." USDA Farmers' Bulletin #2163.

USDA. *A Time to Act*. Washington, D.C.: USDA, January 1998.

USDA Research Service. *Red Meat Yearbook*. Beltsville, Md.: USDA Research Service, 1997.

Vandertuin, Jan. "Vegetables for All." In *Basic Formula to Create Community Supported Agriculture*, ed. Robyn Van En. Great Barrington, Mass., 1988.

Van En, Robyn. "Community Supported Agriculture (CSA) in Perspective." In *For ALL Generations: Making World Agriculture More Sustainable*, ed. J. Patrick Madden and Scott G. Chaplowe. Glendale, Cal.: WSAA, 1997.

Zinn, Howard. *The Zinn Reader: Writings on Disobedience and Democracy*. New York: Seven Stories Press, 1997.

INDEX

G

H

I

S

T